L-15 Scout

L-15 Scout

Boeing's Smallest Airplane

MAL HOLCOMB

McFarland & Company, Inc., Publishers
Jefferson, North Carolina

ISBN (print) 978-1-4766-9285-2
ISBN (ebook) 978-1-4766-5069-2

LIBRARY OF CONGRESS AND BRITISH LIBRARY
CATALOGUING DATA ARE AVAILABLE

Library of Congress Control Number xxxxxxxxxx

© 2024 Mal Holcomb. All rights reserved

*No part of this book may be reproduced or transmitted in any form
or by any means, electronic or mechanical, including photocopying
or recording, or by any information storage and retrieval system,
without permission in writing from the publisher.*

Front cover images: *top*: YL-15s at Boeing ready for delivery (author's collection);
inset Boeing XL-15 sn 46-521 (author's collection); brochure artwork (author's collection)

McFarland & Company, Inc., Publishers
Box 611, Jefferson, North Carolina 28640
www.mcfarlandpub.com

Table of Contents

Preface   1
Introduction   3
Abbreviations   5

1. Post–World War II Outlook for Personal Aviation   7
2. Boeing Looks at a Light Personal Airplane   16
3. XL-15 Liaison Aircraft Competition   29
4. Boeing's Winning Entry   54
5. XL-15 Testing and Design Refinement   64
6. Performance   93
7. Airframe   106
8. Propulsion   126
9. Landing Gear   138
10. Systems   146
11. Weights   175
12. Ground Handling and Maintenance   183
13. Boeing Pilot Comments on the YL-15   194
14. USAF YL-15 Accelerated Service Test   209
15. The Army, the Air Force, and the L-15   217
16. Boeing's Projected Follow-on L-15 Versions   231

Appendix: L-15 Design Patent   237
Chapter Notes   241
Bibliography   249
Index   253

Preface

My interest in liaison airplanes began as a young teenage Civil Air Patrol Cadet in the 1950s. My CAP Squadron had a Stinson L-5 Sentinel and an Aeronca L-16 Champion, in which I got many flying hours. I also had some CAP flights in a two-control Ercoupe like that which had influenced Boeing's light airplane thoughts. Once, the Army National Guard gave us orientation flights in one of their Cessna L-19 Bird Dogs. At age 16, I soloed in a Piper J-3 Cub, the civilian version of the L-4 Grasshopper.

In 1968, I took a position as an aerodynamics engineer working with the Preliminary Design (PD) Group at Beech Aircraft. Two of the old graybeards in the Beech PD Group were Earl Weining and Bill Pierce, who had been key members of Boeing's L-15 engineering team. We all worked for Jack Marinelli, Vice President of Research and Development, a retired Army Colonel. When doing research for this book, I found out that Jack had been heavily involved in the L-15 program when he was an Army officer. Jack had always encouraged my side hobby of historical writing, and we had discussed early Army aviation and liaison airplanes many times, but the L-15 had never come up.

Another of my co-workers at Beech was Bill Brown. Years later, after we had both retired from Beech, Bill e-mailed that he had Earl Weining's L-15 files and thought I could put together a magazine article from them. I told Bill if he would send them to me, I would give it a go. Expecting a large manila envelope, I was astonished when a banker's file box was delivered filled with reports, memos, and photos. Upon inventorying the box contents, the idea for a book replaced that of a magazine article. The real surprise was that the project had started not as a military liaison airplane design but as a civilian private plane.

The Weining files only presented the Scout story from the Boeing point of view. My personal files filled out the data needed on general aviation of the period and an overview of liaison aviation. The Army and the Air Force activities and their respective points of view throughout the program were missing. Also lacking were competition details, including the names of the other competitors.

I found a list of the companies that entered the L-15 competition online in old issues of *The Field Artillery Journal* and *Aviation Week*. The Aircraft Engine Historical Society (AEHS) has an online Finding Aid for USAF Engineering Division Records (RG 342) stored at the U.S. National Archives in College Park, Maryland. The AEHS also has microfilm copies from the National Archives of many Wright

Field Test Reports in its library, and the society president, Kimble D. McCutcheon, was kind enough to make copies of the test reports that dealt with the L-15 for me.

Using the AEHS Finding Aid and the National Archives online search tools, I compiled a list of the Army and Air Force documents held at College Park that dealt with the L-15 program. I spent a week at the National Archives College Park campus with this list in hand. With the archive's skilled and helpful professional staff, I accumulated notes on and copies of an enormous number of Army and Air Force documents and photos.

Introduction

As World War II was in its final stages, the major American aircraft companies foresaw massive unemployment and even shutdowns when their military contracts were canceled at the war's end. Many of these companies believed that with the postwar demobilization of the veterans, there would be a booming demand for personal aircraft. Popular magazines and even the United States government predicted most middle-class families would own an automobile for local travel and a light plane for out-of-town trips.

Boeing Wichita, which had built more than 1,600 B-29 Superfortress bombers and 10,000 Stearman trainers during the war, now found itself with a huge empty factory. The vastly reduced Boeing Wichita engineering team, using company funding, began the design of a light personal airplane as a new product. This included wind tunnel testing and even a flying testbed airplane. The design was not like a typical Piper Cub, and its flap and flight controls were more customary for a large transport or bomber.

In early 1946 a Request for Bids to design a Field Artillery Observation Aircraft (or liaison airplane) was sent out to 26 aircraft manufacturers. The bidders had only 30 days to respond to the request. Boeing Wichita was one of five companies that submitted proposals. The request included detailed requirements and points given on how well each requirement was met. All of the competition entries are described and illustrated here. The Piper PA-9 had the highest rating and number of points, but the contract was awarded to Boeing Wichita as the government wanted to ensure that the Boeing factory stayed open. After Boeing was named the competition winner, the company set aside all thoughts of a civilian aircraft.

The contract was for two XL-15 prototype airplanes, and Boeing engineering launched into the detailed design work. Since the competition had only been for 30 days, the actual XL-15, when it flew, looked a lot different from the competition entry. Design refinement and both ground and flight testing are described in detail in the book. Comprehensive descriptions of the airframe, powerplant, systems, ground handling, and maintenance are included. The airplane could operate with wheels, skis, or floats. The Scout could be towed like a glider by another aircraft or operated from a ship using a Brodie system. The differences between the ten later YL-15s and the original two XL-15 prototypes are delineated. Also included are comments from both Boeing and military pilots.

The L-15 was for Army Ground Forces use, but the Air Force had contracting and technical oversight control of the program. With the involvement of both the U.S. Army and the U.S. Air Force in the L-15 program, there were complexities, interservice rivalries, and politics to be worked out. Boeing was used to working with the Air Force and so took their point of view over that of the ultimate customer, the Army Ground Forces.

Although Boeing spent four years refining the L-15 Scout, its guaranteed performance level was never met, and its unit cost was a magnitude greater than other liaison airplanes.

But even after the Army had told Boeing that the L-15 was dead, the company proposed several redesigned versions of the L-15 as liaison airplanes and looked at commercial derivatives of the L-15 aimed at agricultural, utility, and short-haul transport of passengers and freight.

The L-15 Scout was Boeing's smallest airplane, but it was designed using large-airplane philosophy. Its ultimate value was in keeping the Boeing Wichita division in operation, ensuring its availability for the later B-47 Stratojet production program.

Abbreviations

AAF	Army Air Forces
AAHS	American Aviation Historical Society
AEHS	Aircraft Engine Historical Society
AFB	Air Force Base
AFF	Army Field Forces
AGF	Army Ground Forces
AMC	Air Materiel Command
AN	Army Navy
AOA	angle of attack
ASF	Army Service Forces
ATSC	Air Technical Service Command
AVCO	Aviation Corporation
BAC	Boeing Airplane Company
BHP	brake horsepower
BL	butt line
BW	Boeing Wichita
CAA	Civil Aeronautics Administration
CAP	Civil Air Patrol
CAS	calibrated air speed
CG	center of gravity
Convair	Consolidated Vultee Aircraft Corporation
FAA	Federal Aviation Administration
FS	fuselage station
g	acceleration of gravity
GFE	government furnished equipment
hp	horsepower
IAS	indicated air speed
ISA	International Standard Atmosphere
JATO	Jet Assisted Take Off
LG	Ludington-Griswold
LOC	Library of Congress
MAC	mean aerodynamic chord
mph	miles per hour
n	load factor

Abbreviations

NACA	National Advisory Committee for Aeronautics
NARA	National Archives and Records Administration
NMUSAF	National Museum of the United States Air Force
PA	Piper Aircraft
PD	preliminary design
RFB	request for bid
rpm	revolutions per minute
TAS	true air speed
USAF	United States Air Force
WL	water line
WS	wing station
WSU	Wichita State University

1

Post–World War II Outlook for Personal Aviation

Postwar Predictions

As the apparent end of World War II approached in late 1944 and early 1945, the major aircraft companies became worried about the significant production cutbacks they knew would be coming when the war ended. Countering this was the optimistic expectation of a postwar boom and changes it would bring to the American culture.

Popular magazines were all talking about the way life would be changing in the upcoming postwar era. They predicted that most families of medium or high income would be getting private airplanes to supplement their family automobiles for transportation. Cars would only be for local travel, with planes being for trips and vacations. A survey by the publisher of *Collier's* magazine indicated that 300,000 urban families of medium or high income listed an airplane as their number one or number two priority for postwar purchase. Of the 10 million families responding to the *Collier's* survey, one million stated they expected to buy an airplane after the war. The majority of the writers of the era stated the principal requirement for a postwar private plane was for a cheap, safe airplane with simplified controls.[1]

A contemporary government economy report stated that an ultra-conservative estimate was the number of civil aircraft would increase at a 30 percent annual rate, resulting in more than 400,000 civil airplanes forecast to be in use in the United States by 1955.[2] That same government report said that a more realistic prediction would be a total registration of 1,120,000 family-size airplanes in 1955.[3] It was expected that 97 percent of the military fliers would buy private planes after they finished their wartime service.[4]

This outlook was great news for the small personal aircraft manufacturers like Piper, Aeronca, Taylorcraft, and Cessna, who could once again produce private airplanes. The larger aircraft companies also looked at this potentially huge market as a way to keep their factories open, and most of them started design work on private aircraft projects.

Large Military Manufacturers Consider Private Planes

Republic designed a four-seat, strut-braced high-wing amphibian powered by a Franklin engine. The first prototype, called the Thunderbolt Amphibian, flew in November 1944. After going through some configuration changes, gaining increased

A 1944 magazine advertisement on what postwar family travel would be like (author's collection).

Republic Seabee amphibian (Bill Larkins on VisualHunt).

power, and being renamed the Seabee, it received its CAA type certificate on October 15, 1946, with 1,050 being produced in the postwar period. The Seabee was designed for low-cost production, but Republic, with its big aircraft company labor rate and overhead, lost money on every one of them. With the personal aircraft market falling and the need for production space for the F-84 Thunderjet, Republic on October 4, 1947, announced the shutdown of the Seabee program.[5]

North American designed and produced a four-place, low-wing, retractable tricycle landing gear, single-engine personal airplane called the Navion that looked a lot like the P-51 Mustang. The prototype made its first flight on January 15, 1946, and production was underway by the middle of that year. Then on April 14, 1947, North American announced, "Orders for the Navion were insufficient to justify its manufacture." About 800 aircraft had been delivered, but over 300 Navions were in unsold inventory. North American found itself losing money on every Navion built and saw no way of ever breaking even. In June 1947, Ryan Aeronautical, which had a background in small airplanes, bought the North American Navion design and tooling. Ryan put the Navion back into production and managed to make a profit when producing them.[6]

In May 1945, Douglas set up a project group to design the Model 1015 Cloudster II, a streamlined five-place low wing airplane with retractable tricycle landing gear. It was similar to the XB-42 Mixmaster bomber in configuration and was powered by two Continental 250 hp air-cooled piston engines mounted in the fuselage behind the cabin, driving through a gearbox and extension shaft a single pusher propeller mounted behind the empennage. In late 1947, the project was canceled when it was realized that much development was still needed and that it was financially unsound.[7]

Lockheed also saw possibilities in the expected postwar personal aircraft market and flew prototypes of two potential airplane models, the Model 33 Little Dipper and the Model 34 Big Dipper. The Little Dipper was a single-seat tractor airplane powered by a single Franklin two-cylinder 50 hp engine. The Big Dipper was a two-seat pusher airplane powered by a single fan-cooled Continental four-cylinder 100 hp engine mounted behind the side-by-side cabin driving a pusher propeller.

Top: **North American Navion, which looked a lot like a P-51 Mustang (NARA).** *Bottom:* **Douglas Cloudster II (AAHS—Drew Taylor).**

John Thorp designed both the Dippers. They were both all-metal low-wing monoplanes with fixed tricycle landing gear, constant chord wings with flaps and ailerons, and constant chord horizontal tails of the all-flying stabilator type. These airframe characteristics would show up again in the 1960s in the Piper PA-28 Cherokee, which John Thorp also designed. The first flight of the Little Dipper was in August 1944, while that of the Big Dipper was on December 10, 1945. The Little Dipper, described

Top: Lockheed Little Dipper (Johan Visschedijk Collection, 1000aircraftphotos.com).
Bottom: Lockheed Big Dipper (Johan Visschedijk Collection, 1000aircraftphotos.com).

as a flying motorcycle, had good handling characteristics and performance but would never see production as no market developed for civilian single-seat airplanes. The Big Dipper's performance was better than predicted, and it had good handling characteristics, except for a root stall problem which was being worked on when the prototype was destroyed and the two-man test crew injured in a flight test accident on February 6, 1946. Lockheed canceled the Big Dipper after this accident.[8]

In 1943 Grumman started development on two types of light personal airplanes for the postwar era. The first was the G-63 Kitten, a side-by-side two-place all-metal monoplane with retractable tail wheel landing gear that had lines similar to the F4F Wildcat fighter. Powered by a 125 hp Lycoming O-290, its first flight was on March 18, 1944. The design evolved into the G-72 Kitten II with an extended nose and a tricycle landing gear, making its first flight on February 4, 1946. The G-72 was tested with a single tail like the G-63 and with a twin tail. The twin tail G-72 did away with the rudder pedals and had a two-control system like the Ercoupe with the rudders interconnected to the ailerons. Grumman was fortunate to get many postwar military contracts and, given the slowdown in the light plane market, decided not to go ahead with the Kitten.[9]

Grumman was also an experienced amphibian manufacturer and designed the G-65 Tadpole two-place all-metal amphibian with a 125 hp Continental C-125 mounted as a pusher in a pod above and rear of the cabin. The first flight was on December 7, 1944. After seeing the fate of the Republic Seabee, Grumman decided to stop the Tadpole program. But in the early 1950s Tadpole designer David Thurston went on to develop the Colonial Skimmer based on the Tadpole's layout.[10]

Convair had a light-plane subsidiary, Stinson, which manufactured the popular Stinson Voyager and Flying Station Wagon airplanes, but the parent company

The Grumman Kitten looked like a baby Wildcat fighter (author's collection).

Grumman Kitten II (AAHS).

also developed some airplanes for the postwar market. With the Model 103 Skycar, Convair attempted a simple-to-fly safe airplane designed by George Spratt using the controllable wing concept. The plane had no elevator, rudder, or ailerons. The only controls were a steering wheel, throttle, and brake pedal. The wing was mounted on a pivot over the CG of the aircraft and connected to the steering wheel. When you rotated the steering wheel right or left, the wing pivoted in a banking motion relative to the fuselage, and the front landing gear wheels turned. When you moved the steering wheel fore or aft, the wing incidence relative to the fuselage changed. Thus, the fuselage was always level and unbanked. There was one flying prototype with many articles about it in the press but with little data. After a few years of development, the Model 103 Skycar project was dropped by Convair with no fanfare.[11]

The ConvAirCar was a flying car concept with a conventional-looking automobile that for flight attached to a separate all-metal air unit consisting of a wing, tractor piston engine nacelle, and boom-mounted empennage. The automobile and the air unit each had a powerplant. The automobile was basically a standard car except for special shock absorbers to take landing loads and attachment points for the air unit. The air unit was attached to the top of the car through three load-carrying fittings plus self-connecting flight and throttle control hookups. The initial design was the Model 116, which consisted of a two-place auto powered by a 26 hp engine and an air unit powered by a 90 hp Franklin 4A4 engine. The Model 116 first flew on July 12, 1946, and completed 66 test flights. This was followed by the Model 118, which consisted of a four-place fiberglass body car powered by a 26 hp Crosby engine and an air unit powered by a 190 hp Lycoming O-435C engine. The first flight of the Model 118 was on November 15, 1947. Three days later, the Model 118 crashed when it ran out of fuel on takeoff and destroyed the car, but not the air unit. The pilot had checked the automobile's fuel gauge, but not the air unit's fuel gauge. On January 29, 1948, the Model 118 air unit mated to a new car unit returned to flight testing. An unusual aspect of the ConvAirCar was the automobile and air unit came separately. The car, based on a projected production of 160,000 units, would sell for $1500. You would rent the air unit like a U–Haul trailer. You drove to an airport, rented the air unit,

Top: Grumman Tadpole amphibian (AAHS). *Bottom:* Convair Skycar, which used the Spratt controllable wing concept (Johan Visschedijk Collection, 1000aircraftphotos.com).

1. Post–World War II Outlook for Personal Aviation 15

ConvAirCar, a car with detachable wings and flight engine (Johan Visschedijk Collection, 1000aircraftphotos.com).

and then just left it at the rental office after flying to the destination airport. Like the Skycar, the ConvAirCar project was eventually shut down without fanfare.[12]

In this period of expecting a new postwar personal flying age to blossom, many other companies were working on potential light airplanes, including Boeing.

2

Boeing Looks at a Light Personal Airplane

Boeing Looks at a Light Plane

In early 1945, the trainer aircraft section of the Wichita division of Boeing began a feasibility investigation into developing a simplified light personal airplane for the average person. The first phase of Boeing's research was a theoretical and empirical study to develop a two-control flight control system that would provide an experience more like that of driving an automobile.[1]

Writers of the popular press alluded that the conventional light airplane's number and range of controls were too great for the average person to comprehend. The directional control, in particular, was awkward to learn and use. A longitudinal control capable of stalling the airplane was considered dangerous as it could result in the airplane spinning. Assuming these ideas were correct, it was necessary to investigate the design of an aircraft that eliminated as many of the bad features as possible. Also, eliminating the bad features should not interfere with the airplane's utility, such as low landing speed, good performance, or the ability to operate under adverse conditions.[2]

Simplification of directional and lateral control had been solved in a variety of ways in aircraft such as the Ercoupe, General Skyfarer, and Stearman-Hammond where it appeared to have had no dangerous effects other than some uncertainty in crosswind landings, which required the use of a tricycle landing gear. Compared to the then conventional tail wheel landing gear, the tricycle gear was heavier and more expensive, but it resulted in good landing characteristics, good ground visibility, and ease of taxiing to redeem it. The probability of ground looping was also reduced with tricycle gear.[3]

Restricting elevator deflection had been used more or less successfully to limit the attainable stall angle to a value so that a full stall would not be attained, which could prevent spinning. But there are times when the restricted elevator control authority might be a handicap with an extreme forward center of gravity. A more innovative solution would be to permit the stall to be developed but provide lateral control and spin resistance through the stall. This usually involved wing twist, inboard stall strips, or other devices to ensure that the stall was well developed on the center section before the outer wing panels started to stall. When the outer wing panels are not stalled, the airplane has a good chance of remaining under control and right side up.[4]

2. Boeing Looks at a Light Personal Airplane

General Skyfarer, two-control airplane (AAHS).

The simplified personal airplane must be designed to do many other things besides resisting spins. One of these is the landing approach or glide control. In Boeing's big airplane thinking, almost any pilot could handle approach control by the intermittent use of power if his airplane was not exceptionally clean. Still, in case of power failure, the approach would become dangerous, and some other method would be required, such as flaps, spoilers, or both. In other words, enough glide control needs to be available to permit an approach without the danger of overshooting. An air brake operated by an action similar to that of an automobile brake would be desirable.[5]

The difficulties mentioned in the preceding paragraphs presented a considerable problem that would have to be solved by the Boeing engineers. From their examination of the overall problem, it appeared to require the following controls in the cockpit: elevator, rudder, aileron, throttle, flap, wheel brakes, lift spoiler, and elevator trim tab, not to mention such things as radio, propeller pitch, fuel selector, etc. Boeing was used to designing for military and airline pilots and held a somewhat low opinion of personal aircraft pilots. Boeing's looking down at non-professional pilots is shown in this comment by James M. Wickham, Chief Aerodynamicist, Boeing Wichita: "It was perfectly obvious that this list represented no control simplification for the average, and presumably stupid, private pilot."[6]

Stearman-Hammond Y-1, two-control airplane (Schiphol Airport Archive).

Now all of the control functions listed were useful, and the complete elimination of the role of any one control was not desirable. The obvious approach was to combine as many functions in as small a number of cockpit control elements as possible. The automobile arrangement of steering wheel, brake pedal, and throttle was considered the ideal arrangement of the primary cockpit controls from the simplified flying standpoint. It was felt the throttle, for fine adjustment, should be the hand type, not a foot pedal, and the steering wheel should have fore and aft motion to provide control of pitch. Adding the air brake spoiler control to the brake pedal was an obvious solution to eliminate a separate spoiler control. The pilot would merely push the brake pedal in the air if he wished to steepen his glide. This action would be natural and similar to the action of using brakes on the ground. The application of wheel brakes in the air does no harm, while the action of the air brake (lift dump) spoilers on the ground is an aid to the wheel brakes in obtaining traction at speeds just below landing speed by reducing lift and would also provide more wind stability when the airplane is parked with brakes on. Spoilers are a potentially powerful control, and flight testing would be required to determine the optimum size and configuration. Spoilers are more effective on a clean airplane and less so on a dirty, or heavily flapped, airplane.[7]

Although wing flaps were not used on many small personal airplanes in that era, they are beneficial to achieve high lift coefficients for landing. Since short landing and takeoff ability is an important factor in the utility of an airplane, high lift coefficients as provided by flaps were assumed to be a requirement. Since the flap control is merely another means of adjusting lift, it was considered logical to combine the flap and elevator control and allow the fore and aft motion of the wheel to actuate both surfaces. A further thought was to eliminate the elevators and use the flaps as the only lift control. This idea required a close examination, since elevators had always been the most primary and fundamental control on an airplane. The elevator is an attitude control that varies the attitude (angle of attack) of the airplane with respect to the air stream, and as a result, varies the lift coefficient of the wing. While the wing flap, on the other hand, varies the lift coefficient by an effective wing camber change, and the attitude or angle of attack remains relatively constant. Thus, the flight control system should result in a constant angle of attack relative to the flight path, and the pilot would be able to select a nose position relative to the horizon that he found most desirable. Large attitude changes resulting from flap action are generally offset by tail design and center of gravity location. The end result of both flap and elevator action is a change in lift coefficient, and it is this change in lift coefficient which controls the airplane, not the action of the flaps or elevators in themselves. Thus, it is reasonable to assume that longitudinal control could be obtained utilizing flaps alone, and the elevator control system could be eliminated. An external airfoil-type flap suggests itself for this application, since it is mechanically simple and is capable of both positive and negative deflections.[8] The Boeing Wichita engineers had used an external airfoil flap on the prewar Stearman XOSS-1 Navy observation prototype and were somewhat familiar with their characteristics.

The response of a flap-type longitudinal control would be expected to be somewhat faster than an elevator response due to the mass inertia of the airplane

Boeing Wichita engineers had used external flaps on the prewar Stearman XOSS-1 (NARA).

preventing an instantaneous change of attitude. Also, climb-out would occur with a lower nose attitude than in the case of no flaps and thus would provide better visibility for the pilot. The attitude changes with airspeed should be small, with the actual variation controlled by the size and location of the horizontal stabilizer and center of gravity. The most apparent disadvantage inherent in this proposed system is the lack of trim control to overcome the ground effect in the last stages of the landing flare. The seriousness and solution of this problem would have to be determined from flight tests and would require development testing on an experimental airplane. Using a stabilizer trim tab and very moderate flap angles on approach could provide the necessary flare margin. From the standpoint of ground effect, a high location for the horizontal tail or a V-tail would be advantageous.[9]

A longitudinal trim tab would be needed in any case to adjust for a shift in the center of gravity and for cruising attitude control. The trim tab could also be used as an emergency longitudinal control in case of failure in the flap control system. The proper use of the lift spoilers in conjunction with the throttle could be a third longitudinal control in an emergency.[10]

Boeing's hypothetical simplified private airplane had flaps for low landing speed and spoilers for glide control, combined in a two-control airplane. The lateral-directional control system should have two independently operating elements to provide control in case one element fails. This could be a dual aileron system or a combination of aileron and rudder. Differential spoilers could also be used, and by properly sequencing the spoiler deflection and the differential aileron motion of the external-airfoil flaps, it appeared that the rudder could be eliminated as the yawing characteristics due to the spoilers and the differential flaps offset each other.[11]

There were two-control aircraft in production, such as the Ercoupe, General Skyfarer, and Stearman-Hammond, but they did not have the pitch characteristics of this Boeing design study or the Convair Skycar. The Ercoupe eliminated the rudder pedals but still had rudders geared to the ailerons. The Skyfarer and Stearman-Hammond did not have a rudder control system and used ailerons alone to turn.

Convair with its Spratt control wing Model 103 Skycar had a similar longitudinal flight path; however, the Skycar used a pivoting wing with no moveable surfaces instead of the Boeing concept using a fixed wing with moveable control surfaces. While both designs had no rudder system, they had different lateral-directional flight paths. The Boeing design would bank in a turn like a conventional airplane, while the Convair Skycar tilted its wing and the fuselage remained straight (i.e., the wing banked, but the fuselage did not).[12]

Since the Boeing design was radically different from existing aircraft designs, the Boeing engineering team realized that designing a simplified personal airplane would require considerable development work and experimental testing to verify the concept. They also realized that the proposed longitudinal control system might have features that make it objectionable to a pilot. Boeing Wichita decided some preliminary experimental work, both wind tunnel and flight testing, needed to be made to investigate the problems of the simplified control in general and the flap longitudinal control in particular.[13]

The preliminary analytical studies indicated that the proposed simplified flight

2. Boeing Looks at a Light Personal Airplane

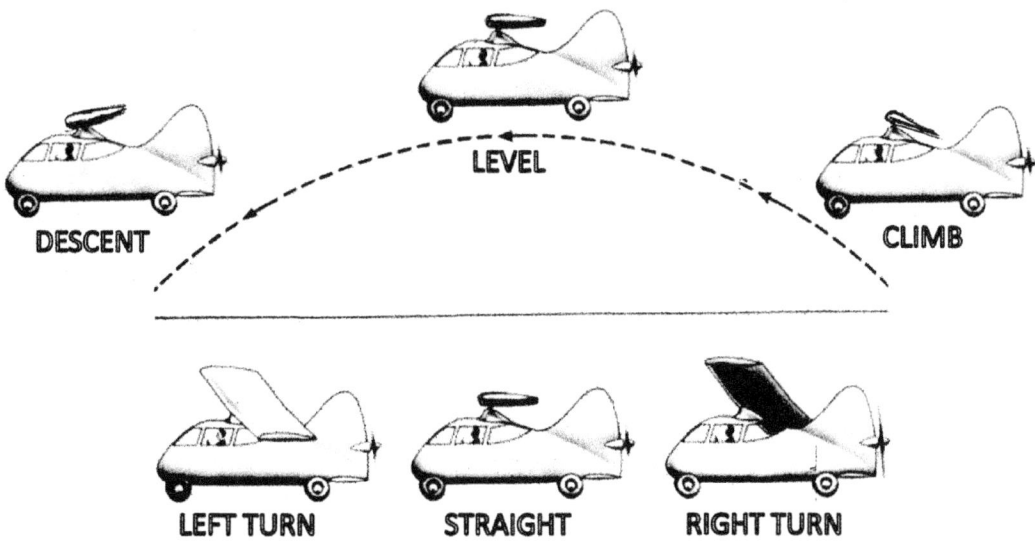

The Convair Skycar, with its Spratt control wing, had a longitudinal flight path similar to what Boeing was pursuing for their light plane design (author's collection).

control system was feasible and that a CG location somewhat aft of that normal for conventional airplanes should be coupled with a larger than normal horizontal tail to provide the best handling characteristics. Because of the large size of the flaps compared to the size of ordinary elevators, it was necessary to balance the hinge moments rather accurately to provide reasonable control forces for the pilot. The available hinge moment information on the external airfoil type of flap was minimal. It was decided to run two-dimensional wind tunnel tests on a Boeing 117 airfoil section with an external airfoil flap to determine the variations in characteristics as affected by changes in flap size, the position of the flap nose relative to the wing trailing edge, and hinge locations. In particular, the flap hinge moments at negative deflections were unknown and negative deflections were necessary for the proposed control system. Tests were made on several hinge locations for each flap size, and tests with negative deflections up to −20° were run. A total of 117 test runs were made in the University of Wichita four-foot wind tunnel during the latter part of 1944. Flaps of two-inch chord and three-inch chord were tested on an airfoil with a ten-inch chord. The lift, drag, and pitching moments of the airfoil and flap combinations were measured, as well as the flap hinge moments.[14]

The wind tunnel test results indicated that it would be reasonable to start further work using a 20 percent chord flap. The maximum lift coefficient of the combination in the wind tunnel was 2.19, which was calculated to be 2.40 for a full-scale light airplane at landing speed. These coefficients were based on the total area of the wing plus flap. The flap hinge moments for deflections from +40° (trailing edge down) to −10° (trailing edge up) were small and stable. Beyond −10°, the hinge moments were erratic but in no case reversed, but also had minimal control effect, making it probable that negative, or trailing edge, up flap deflections should be limited to –10° on the airplane.[15]

L-6 Testbed

No Boeing airplanes were of a size or configuration that could be modified to be the test aircraft. So, an Army surplus Interstate L-6 liaison airplane was obtained to be modified to incorporate the flap control and spoiler for flight test evaluation. Before the flight tests, three-dimensional wind tunnel model tests of the baseline L-6 and the modified L-6 were run at the University of Wichita. The three-dimensional wind tunnel test results were analyzed along with those of the two-dimensional flap section tests to give a good general picture of the control forces and airplane attitudes that would be expected in the full-scale flight test.[16]

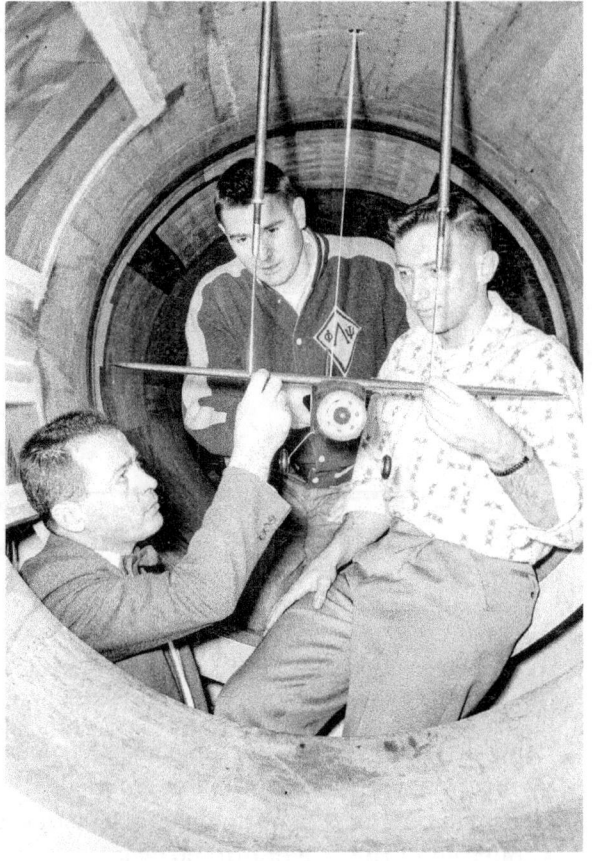

Boeing tested the L-15 at the University of Wichita wind tunnel. The wind tunnel is shown here with a Cessna model (photograph by Don Dalrymple, Wichita State University Libraries, Special Collections and University Archives).

The $\frac{1}{10}$ scale model used for the wind tunnel tests was not an exact reproduction of the L-6 airplane or its modifications because no detailed L-6 drawings were available, and the new wing and flap combination had not been designed at the time the wind tunnel model was built. The L-6 wind tunnel model shapes were approximated from a small three-view drawing and measurements of the aircraft. In addition to the full-span external airfoil flaps, the wind tunnel model tested upper surface spoilers of the split flap type to evaluate for glide control and aileron control. To use the original L-6 tail, the wing area of the modified airplane was reduced below that of the standard L-6.[17]

The wind tunnel results indicated that the elevator control forces should be small and stable over most of their range. The airplane attitude change with flap angle should be small and negative for most CG locations likely to be used. This indicated that the horizontal tail size was smaller than optimum for the type of longitudinal control being considered, but the static longitudinal stability was expected to be acceptable.[18]

Boeing obtained an Interstate L-6 like this to modify as a testbed for their personal airplane studies (NMUSAF).

Wind tunnel rolling and yawing moment results due to differential flap travel indicated that the plus or minus 5° of flap travel proposed for aileron control would give rolling moment coefficients up to 0.025, which was smaller than desired but assumed to be safe for flight test. The adverse aileron yaw was not as bad as had been expected.[19]

The wind tunnel test results on the glide control spoilers showed them to be quite satisfactory, and at a lift coefficient of 0.6 with 0° flaps, the lift-drag ratio of the model was reduced from 9.3 to 4.9. Trim and attitude changes due to the spoilers were negligible.[20]

When the upper surface split flap–type spoilers were tested for aileron control, they were unsatisfactory. The principal problem was a reversal of effect at small spoiler deflections. A small deflection intended to give a right roll would give a left roll instead. At larger deflections, the roll was in the correct sense.[21]

There was no ground effect wind tunnel test run to examine the nature of the nosing down tendency expected at the end of the landing flare. This test was omitted because the Boeing engineers thought wind tunnel ground plane tests were not particularly accurate, and the problem was affected by airplane mass inertia in addition to aerodynamic forces. They felt flight technique and pilot reaction would be the deciding factors rather than a numerical increase in stability near the ground.[22]

There were no showstoppers in the wind tunnel results, so the design and construction of the L-6 modifications began. The strut-braced test wing was plywood covered with glide control spoilers. The test wing had a span of 30 feet with a 50-inch chord, and attached to that was a full-span 10-inch-chord external airfoil flap. A new longitudinal control system permitted the pilot to fly the airplane with elevators or flaps, or both simultaneously.[23]

While the test modifications were being designed and built, baseline flight testing was being done on the standard Interstate L-6 airplane to measure control forces and flight characteristics as a reference basis for evaluating the new system. The test pilot had a good impression of the stability and control of the baseline Interstate L-6, and thus these characteristics were assumed to be a good standard toward which to gauge the unconventional system against.[24]

Flight testing was done to determine the possibility of using either rudder or ailerons alone for the turning control. Measurements of turn rate were made for various control applications, and the skid or slip was measured on a yaw meter. Rudder alone was determined to be a poor turn control, and although ailerons alone were much better, they were not perfect. It was believed that a larger vertical fin that would limit the side slipping might make the ailerons a good turn control.[25]

Neighboring Cessna Aircraft had obtained an Ercoupe for competitive evaluation. Boeing was able to obtain the use of this Ercoupe from Cessna for flight evaluation of this certified two-control airplane. The Ercoupe's longitudinal stability and control and rate of roll were measured. Several Boeing pilots flew the Ercoupe to become acquainted with the operation of a two-control airplane. The pilots' opinions were rather noncommittal, but it was concluded that the operation was very simple and that taxiing, in particular, was a significant improvement over a tail wheel–type airplane. The Boeing pilots commented that crosswind landings seemed to induce a helpless feeling due to lack of rudder. A Boeing paper on the Ercoupe testing noted that all the Boeing pilots were used to three-control airplanes, which might explain some of their difficulty. The Boeing pilots felt that the Ercoupe had a slight wallowing tendency in rough air, which could be objectionable on a long flight. This particular Ercoupe had seen rough use, and the control system was worn and sloppy to the extent that could easily account for some of the wallowing encountered. The Ercoupe's longitudinal stability was good, and the rate of roll was quite adequate, with automatic turn coordination better than a three-control pilot usually gets. The Ercoupe's hands-off spiral stability was poor, but it was noted that likely the light control forces coupled with the friction in the system permitted the controls to become stuck in a turning position, thus causing a turn to wind up rapidly. Stalling under all conditions was very good, with lateral control available at all times. It was commented that the Ercoupe's flight characteristics were such as to make it almost impossible to get hurt other than by flying into a tree or fence.[26]

Modification of the L-6 with the test wing and controls was completed in the latter part of 1945. Familiarization flights and rigging checks were made, and then the actual flight tests began. The speed-power, stability, and roll characteristics of the modified airplane were determined from the flight tests. The center of gravity was progressively moved aft to a point directly below the airplane's aerodynamic center. This was done to minimize the effect of flap position on longitudinal trim. The height of the aerodynamic center above the center of gravity was sufficient to provide good stability.[27]

The first problem encountered in the modified L-6 flight test was poor roll rate and aileron overbalance (light forces). The external airfoil flaps could be depressed 40° to the wing chord line or rotated 10° above the chord line. The flaps could operate differentially, and it took approximately 4° flap movement for aileron control. The low roll rate was somewhat expected, but the overbalance was not. Luckily, a solution to both problems was made by adding a 1.5-inch metal strip trailing edge extension to the flap's full span, which resulted in an improved roll and aileron feel. Due to the flap trailing edge extension, the control force required to move the flaps in place of the elevator as longitudinal control increased more than that desirable for deflections beyond 20°. The flap used as a longitudinal control was evaluated in flight and on landings. The control was marginal for landings on the L-6, which had a tail wheel landing gear. Some landings flared smoothly using the flaps for longitudinal control, and others required some input from the actual elevators. The effect of the flap as a flight path control was much more rapid than that of a conventional elevator control. This was because a conventional airplane with an elevator obtains a flight path change due to an attitude change of the airplane, which results in a change in wing angle of attack and therefore lift. This requires that the inertia of the airplane be overcome with a resulting lag in the control effect. The modified L-6 obtained its flight path change through an immediate increase or decrease in lift resulting from a change in flap deflection. Since the rotational inertia of the airplane does not have to be overcome to change flight path angle, there is no control lag.[28]

The experimental control system in the modified L-6 was a pilot's nightmare,

Boeing borrowed an Ercoupe similar to this one from Cessna to get some two-control flight experience (Wiltshirespotter on VisualHunt).

and it was decided to hook the flaps into the regular stick control and put the elevators on an auxiliary control similar to a trim tab. At this time, 1° dihedral was also removed from the wings to further aid lateral control. After the flap was connected to the normal control stick, the sensitivity and control forces were still much too great, and two successive changes were made to reduce them to acceptable values.[29]

The glide control spoilers were tested in flight and found to work satisfactorily, giving increased sink with no trim difficulties. However, the spoilers were found to be of little or no value to the test airplane due to the inherent high rate of sink for the configuration.[30]

At this point in the flight tests, the test airplane was capable of ordinary flight control by use of flaps only but could not make satisfactory landings without help from the elevators. This was due in part to the ground effect mentioned previously and to the particular configuration and center of gravity of the test airplane. The L-6 test airplane had a rather small wing area, moderate aspect ratio, and high drag. This resulted in a high rate of sink which was not ideal for the demonstration of the simplified control system.[31]

The following is a comment by Boeing test pilot Elton Rowley of a landing in the L-6 testbed: "Holding the previously mentioned 70 miles per hour approach speed and trim at 5 or 6 degrees up elevator; upon coming close to the ground where flaring of the glide was necessary, the flaps were deflected downward slowly. During this flap application the ship decelerated and the tail started to rise. Just before contact with the ground was made, the stick was pulled full back, giving full flap deflection. The speed at this point was probably 58 to 60 miles an hour. Contact was made with the ground in a very tail high attitude. Still determined not to use elevator to bring the tail down but to find out what was necessary without the use of elevator to recover from this position, power was applied. This very definitely aggravated the already bad situation. The throttle was immediately closed and the flaps raised; the tail immediately came down. The propeller was not damaged but it is positive that it would be if extreme care was not used."[32]

The test pilot's comments on the landing clearly pointed out the need for a configuration change to the airplane. The change was to move the horizontal tail 29.5 inches up on the vertical fin to a mid-tail position. This repositioning of the horizontal tail improved the pitch dampening at cruise and high speed and made some improvement in the landings. The landings, however, were still far from what they should be. A further change was made by depressing the propeller thrust line 3°, so the thrust line went through the CG to eliminate the power effect due to thrust line offset. But flight tests showed only very minor improvement in longitudinal trim due to this thrust line change.[33]

The spoilers on the testbed L-6 aircraft had not been set up to differentially operate as ailerons, thus flight testing to determine the feasibility of eliminating the rudder control was not made.[34]

Only a few further tests were made before flight testing was discontinued due to the poor condition of the engine and the advent of the XL-15 program. The L-6 test airplane was later used to support the flap and aileron development as well as the propulsion system development for the XL-15.[35]

Boeing's look at a light personal owner airplane effectively stopped at this point due to manpower and resources needed for higher priority programs. Most of the manpower working on the personal aircraft project was shifted to the XL-15 liaison aircraft program. Also, with the new B-47 jet bomber coming into the picture, there was no need to manufacture a private plane to keep the Boeing Wichita factory doors open.

Boeing's Closing Thoughts on a Personal Airplane

Although the Boeing simplified control personal airplane was essentially dead, the engineers wrote some thoughts on what should be done if the program should ever get restarted. They realized that the simplified flight control system was still a long way from being developed but believed much useful information had been obtained in the wind tunnel and flight testing. Many problems had been encountered or identified, but the only severe ones were the strong increase in stability near the ground coupled with the normal increase in stability of a high wing airplane at low speed. This stability increase resulted in the pilot running out of trim control during the landing flare. The high drag of the modified L-6 test aircraft added to the problem of the high sink rate, resulting in a minimal choice of acceptable power-off approach speeds.[36]

The engineers concluded that any further development of the system should be aimed at eliminating or improving the ground effect problem. This should include wind tunnel and flight testing of a configuration specifically designed for the proposed simplified flight control system. A completely new experimental airplane would be required for flight development as the L-6 would not do.[37]

The L-6 testbed with the new wing with external airfoil flap and raised horizontal tail (author's collection).

The L-6 testbed was a stick and rudder pedal-controlled aircraft with a conventional tail wheel landing gear. The new aircraft should have a control wheel but no rudder pedals. The airplane should have tricycle landing gear with a steerable nose wheel tied to the control wheel.[38]

The new design should have a high aspect ratio wing for both performance and control reasons. Using a high aspect ratio would allow the flap to have a smaller physical chord for the same percent chord, reducing control forces without using aerodynamic balance. Induced drag, downwash, and ground effect are also less with a higher aspect ratio wing. The aircraft should have low drag, which would provide a flat glide and low rate of sink, permitting spoilers to give very good glide control and would decrease the landing flare troubles. If 20 percent chord flaps are used, the center of gravity should be in the region of 50 percent to 60 percent MAC. The horizontal stabilizer should be located as high as possible to reduce the ground effect and should be large enough to handle this extreme aft center of gravity location. Although the horizontal tail would not have an elevator, it should be an adjustable incidence stabilizer or have a trim tab.[39]

When Boeing started the light plane study, it was believed that the aircraft should be stall proof. However, because of the ground effect problem, it appears that free air stalling would be acceptable and made safe by designing for proper stall progression and spin resistant characteristics. This would make it easier to keep the tail down in the landing flare. The airplane would then have attitude changes similar to but smaller than those of a conventional airplane. Jim Wickham stated the initial studies and tests had all been based on a high wing configuration, but an aerodynamic analysis of the flare leads toward considering a low wing.[40]

However, the Project Engineer, Earl Weining, believed that if the personal aircraft program was restarted, it should be a high wing with minimum CG travel.[41]

The simplest dual lateral-directional control would probably consist of plug or retractable spoiler-type ailerons coupled with a rudder and no flap differential. This would allow either the rudder or the aileron control system to become inoperative and still leave some directional control. Both the rudder and spoiler ailerons would be controlled by the rotation of the control wheel in automotive steering wheel fashion.[42]

The team members concluded that an airplane designed as described in the preceding paragraphs, if adequately developed, would have the low landing speed, good glide control, and simple operation desired for a safe personal aircraft.[43]

As previously noted, the L-15 liaison aircraft competition had come into the picture, and Boeing management decided to discontinue any further studies on the light airplane. Boeing Wichita saw a drop in work and manpower at the end of World War II, but it soon turned upward with the L-15 and B-47 programs coming on. The L-15 program got most of its personnel from the light airplane research program, and some of the results of that program fed into the L-15 design studies. In retrospect looking at the actual postwar light plane picture, the huge production predictions and the expected prevalence of safe simplified airplanes did not become a reality.

3

XL-15 Liaison Aircraft Competition

Army and Air Force Relationship on Liaison Aviation

At this time, the U.S. Army had three major components: the Army Ground Forces (AGF), the Army Air Forces (AAF), and the Army Service Forces (ASF). The AGF and AAF were combat organizations, while the ASF was a support organization. Both AGF and AAF operated liaison airplanes, but they had different philosophies on using them. The AGF had gotten their liaison airplanes in a back-door way when the AAF had shown they could not provide the AGF with the aerial observation services needed by the field artillery.

The Army Air Forces expected to shortly become a separate armed force equal to the Army and Navy. They also felt that they should be the providers to the Army of all aviation services and the Army should not have its own aircraft. However, the road to becoming a separate United States Air Force (USAF) was a very bumpy political path. The AAF needed the political backing of the Army general staff to get their separation; thus, they were agreeable, although reluctantly, to allow the Army Ground Forces to have aircraft.

When the AGF needed items like trucks, artillery, and armored vehicles, they were developed and obtained through the ASF. But liaison airplanes for the AGF were developed and obtained through the AAF, their rival for operating them.

Competition Request

On September 27, 1945, the Engineering Division of the Army Air Forces, the Air Technical Service Command, was ordered to procure two or three experimental flight vehicles that were to meet the "Military Characteristics of Field Artillery Observation Aircraft, dated August 20, 1945."[1]

A Request for Bids (RFB) was sent out to 26 aircraft manufacturers on February 4, 1946, for an Informal Design Competition of Field Artillery Observation Aircraft to select the most suitable model to be developed under an experimental requirement. Proposals were to be received by the Air Technical Service Command (ATSC) within 30 days.[2]

RFB Requirements

A. Performance Requirements

Table 3.1

Requirement	Unit	Minimum	Desired
TO dist over 50 ft—sod	ft	600	300
Landing dist over 50 ft—sod	ft	600	300
Landing Speed	mph	35	25
Max R/C	fpm	600	800
Service Ceiling	ft	15,000	18,000
Cruise at 75% Power	mph	90	–
Range, with aux tank	s. miles	500	–
Endurance, standard fuel	hours	3.5	4

B. Engineering Characteristics

1. Structure and Design:
    a. An airplane having a high strength-weight ratio is desired to permit the versatile utilization of the aircraft.
    b. The maximum Figure of Merit will be rewarded if AAF design criteria are used. CAA regulations will be accepted, however, at a reduction of fifteen (15) points from the maximum Figure of Merit.
    c. Simplicity of design details and operation is desired.
    d. Efficient utilization of space providing practical arrangements of crew in relation to controls and visibility is desired.
    e. Provisions for installation of hook for use of Brodie take-off and landing equipment is desired.
    f. A spin resistant airplane is desired.
    g. Essential safety features such as safety belts, shoulder harnesses, cockpit padding, and emergency exits are desired.
    h. Design of components and parts adaptable for production in quantity is desired.
    i. Provisions for quick interchangeability of ski-type alighting gear for wheel type is desired.
    j. Provisions for quick interchangeability of pontoon-type alighting gear for wheel type is desired.

2. Power Plant Installations:
    a. The use of a reliable engine is desired. (The 125 hp Lycoming O-290-7 is the recommended engine.)
    b. Emergency operation on all-purpose fuel and oil is desired.
    c. Fuel economy in operating the aircraft is desired.
    d. Normal operation in temperate climates and provisions for operation in extreme cold, tropical, dusty, and salty atmospheres is desired.

3. XL-15 Liaison Aircraft Competition

    e. The use of bladder-type fuel cells interchangeable with self-sealing cells is desired.

    f. Due consideration is to be given proper fuel and oil systems, cooling and exhaust systems, and electrical systems, from the standpoint of operational safety.

3. Maintenance and Repair:

    a. Accessibility for inspection, easy and effective maintenance and repair of all major components is desired.

    b. The use of quick disconnect features, including ease and rapidity of engine change, interchangeability of parts, and the elimination of special tools required, is desired.

    c. Lightweight aircraft having minimum overall dimensions to facilitate ground handling is desired.

4. A compact radio installation, readily accessible for operation by both crew members, and rapid replacement, is desired.

5. Standard instruments and instrument panels are desired.

6. Suitable lighting for night flying and navigation is desired.

7. The use of an electric self-starter is desired.

8. A suitable exhaust-type cabin heater is desired for normal usage.

9. Easily removable armor protection for crew against light flak and small arms fire from below and rearward is desired.

10. Provisions for the use of a vertically mounted K-20 camera is desired.

11. Provisions for the installation and use of message pick-up equipment is desired.

C. Suitability Characteristics:

1. Operational Utility:

    a. Operation from small unprepared (muddy and plowed) fields is desired.

    b. Unobstructed visibility for the pilot is desired with emphasis on forward and downward visibility.

    c. Unobstructed visibility for the observer is desired with emphasis for downward and rearward visibility.

    d. A speed of 60 mph for sustained level flight is desired.

    e. Ability to recover from 200 mph dives will be rewarded maximum Figure of Merit in points. The maximum points for Figure of Merit will be deducted if this diving speed equals 150 mph or less.

    f. Provisions enabling the full winterization of this aircraft is desired. Feasibility for efficient operation under all climatic conditions is desired.

    g. The aircraft should be transportable on a standard U.S. Army two and one-half ton cargo truck, with minimum disassembly.

h. The suitability for being towed by another aircraft at a speed of 180 mph will be rewarded the maximum Figure of Merit in points. The maximum points for Figure of Merit will be deducted if this towing speed equals 150 mph or less.

Each submitted design was given a Figure of Merit set of points based on the guidelines as shown in the table.

Table 3.2. Figure of Merit Guidelines

Factor	Max Points
Performance	**200**
Take-off Distance over 50 foot Obstacle	60
Landing Distance over 50 foot Obstacle	40
Landing Speed	40
Rate of Climb	25
Service Ceiling	15
Endurance	10
Ferry Range	5
Cruising Speed	5
Engineering	**500**
Structure and Design	*150*
Strength-Weight Ratio	40
Design Criteria	35
Simplicity of Design	20
Spin Characteristics	16
Arrangement	15
Safety Features	8
Adaptability to Quantity Production	7
Provisions for Brodie System	3
Provisions for Skis	3
Provisions for Pontoons	3
Power Plant Installation	*110*
Choice of 125 HP Engine	25
Fuel Economy	20
All-Weather Operation	20
Self-sealing Tanks Provisions	15
Choice of 150 HP Engine	10
Safety Features	10
Choice of 190 HP Engine	5
Emergency Operation on All-Purpose Fuel and Oil	5
Maintenance and Repair	*165*
Ease of Inspection, Maintenance and Repair	75
Rapid Assembly and Disassembly	50
Ease of Ground Handling and Movement	40
Equipment Installation	*75*
Radio	18
Instruments	14
Night-Flying Equipment	12
Self-starter	10

Factor	Max Points
Cabin Heater	9
Armor	5
Vertical Camera Mount	4
Message Pick-Up Equipment	3
Suitability	**300**
Operational Utility	*300*
Operation from small, unprepared fields	100
Visibility—Pilot	45
Visibility—Observer	30
Speed for sustained level flight	50
All-weather Operations	30
Performance of Evasive Maneuvers	25
Transportability	15
Suitability for being towed by another aircraft	5

For each of the Performance items, points were calculated by an equation similar to the following

$$\text{Points} = \frac{(\text{min acceptable} - \text{proposal value})}{(\text{min acceptable} - \text{desired value})} \times (\text{max points for item})$$

For items in the Engineering categories, the points were determined by committees of Wright Field Air Force engineers. For each of the Suitability characteristics, the points were determined by a committee made up of Wright Field Air Force engineers and by the Army Ground Forces Liaison Officer assigned to the ATSC (soon to be renamed the Air Materiel Command).[3]

Submittals

Five companies submitted eight proposals to the RFB. These were

Bellanca Aircraft Corporation—two proposals
Boeing Aircraft Company (Wichita division)—one proposal
Consolidated Vultee Aircraft Corporation (San Diego division)—two proposals
Ludington-Griswold, Inc.—one proposal
Piper Aircraft Corporation—two proposals

All were paper designs and are described in the following sections.[4]

Bellanca

The Bellanca Aircraft Corporation of New Castle, Delaware, was founded in 1927 and had produced many commercial, personal, and military airplanes. Their only previous experience in the field of Army observation aircraft was the Bellanca

YO-50, a prototype observation aircraft built for the United States Army in 1940. Typical for aircraft of its type, it was a high-wing strut-braced monoplane with tail wheel undercarriage and extensive cabin glazing. The Army purchased three examples of the YO-50 for evaluation against the Stinson YO-49 and Ryan YO-51 Dragonfly. The Stinson O-49 won the production contract, and no further YO-50s were built.

Bellanca submitted two proposals to the Informal Design Competition of Field Artillery Observation Aircraft, for the Bellanca models 20–13 and 24–19. Both were two-place tandem high wing aircraft with tricycle-type landing gear. The two designs were similar in configuration and construction but different sizes.

The wings were of strut-braced monoplane type with a double drag truss. Wing and flap construction consisted of wood spars and ribs with plywood leading edges and fabric covering. The fixed-wing leading-edge slats were constructed of wood spars, ribs, and covering. The wing high-lift systems were based on the YO-50 with fixed leading-edge slats and slotted main flaps that had auxiliary slotted flaps (today, the flaps would be called double-slotted flaps).

The tail sections were constructed with wood spars and ribs and were fabric covered.

The fuselage was of welded tubular chrome molybdenum tubing of conventional truss-type design and covered with fabric. Separate doors were provided for the pilot and observer.

The landing gear was of fixed tricycle configuration with oleo shock absorbers. The main wheels were 7.00 × 6 size with hydraulic brakes operated by the pilot only.

Bellanca YO-50, 1940 prototype observation airplane (author's collection).

3. XL-15 Liaison Aircraft Competition

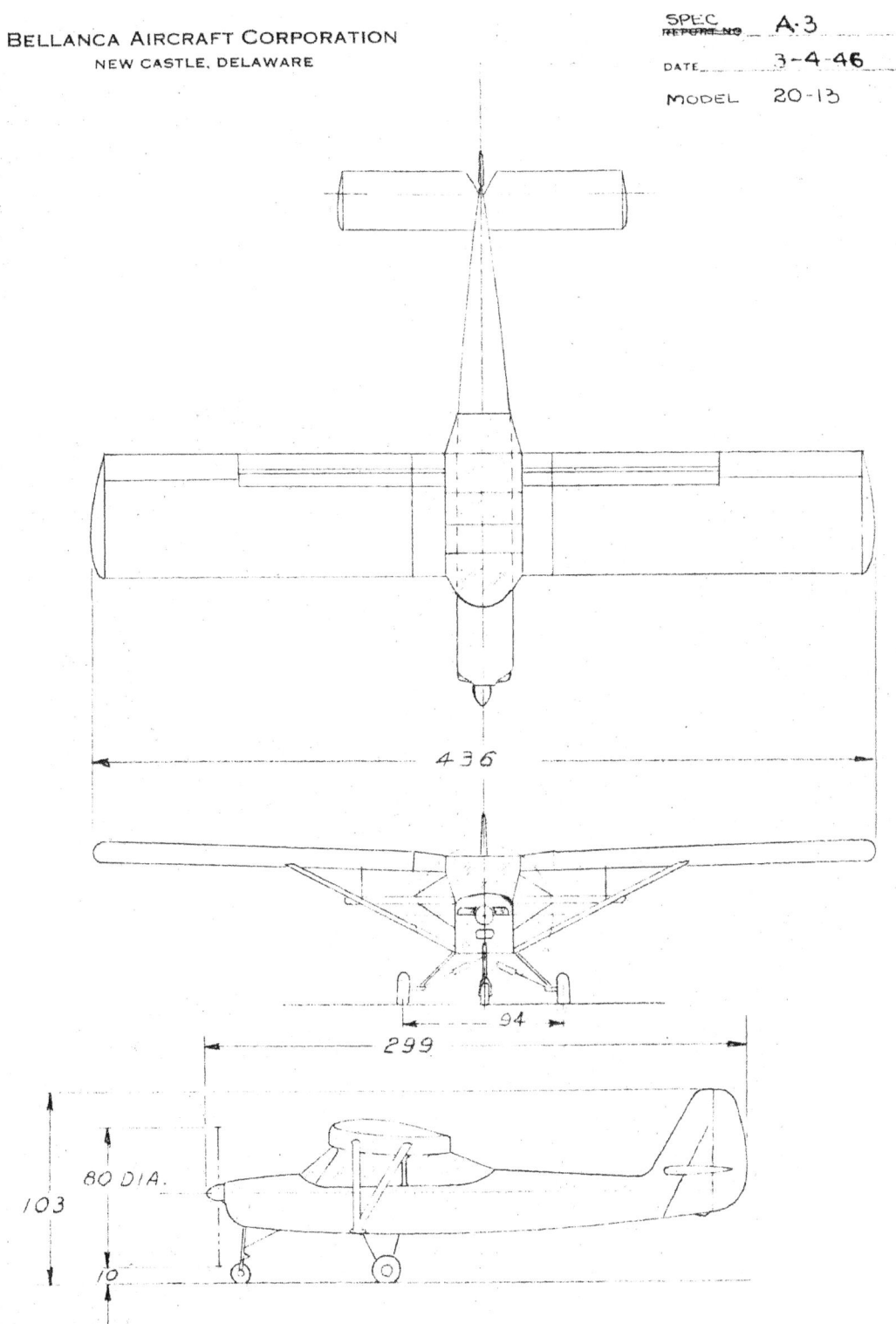

Bellanca Model 20–13 three-view (NARA).

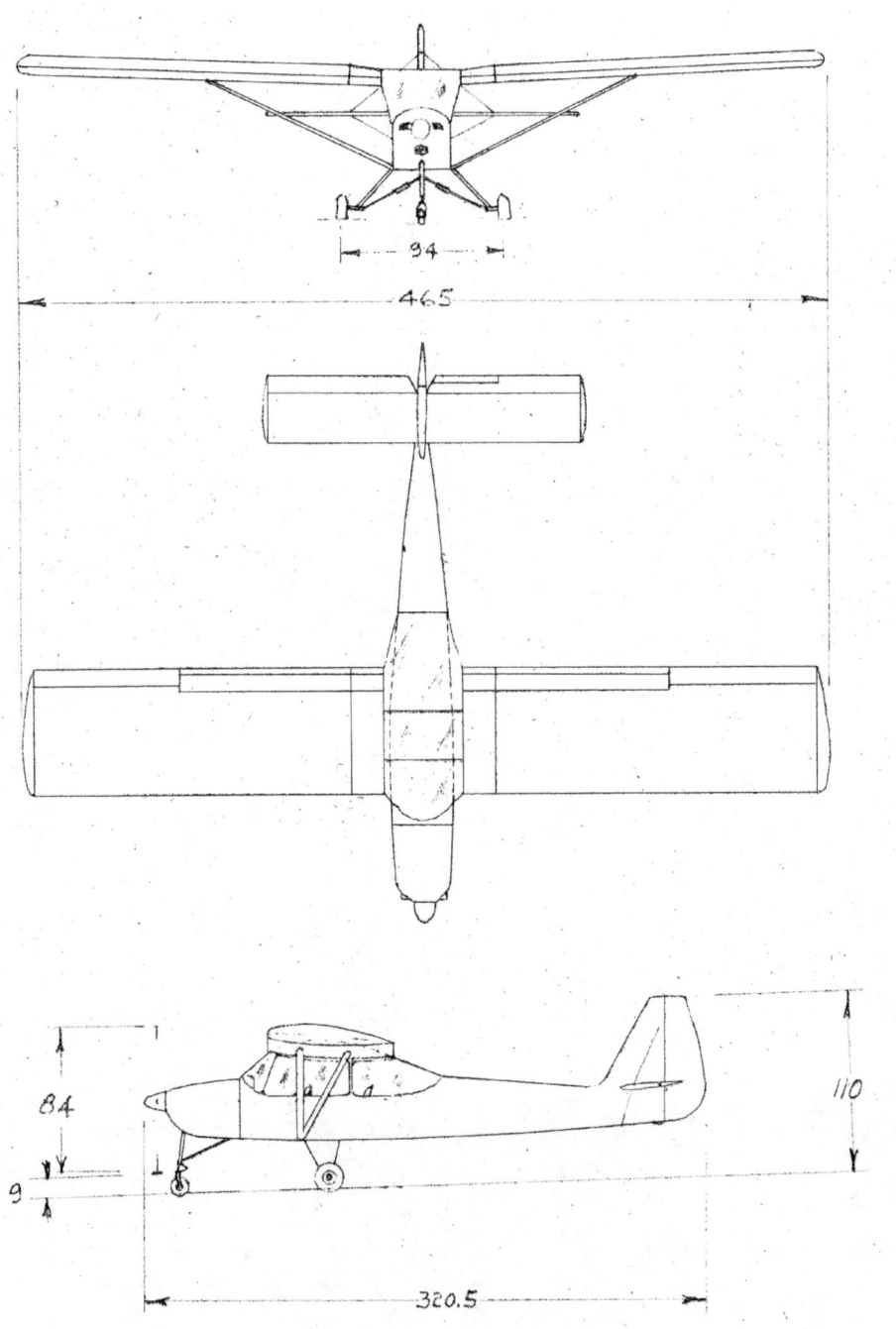

Bellanca Model 24–19 three-view (NARA).

3. XL-15 Liaison Aircraft Competition

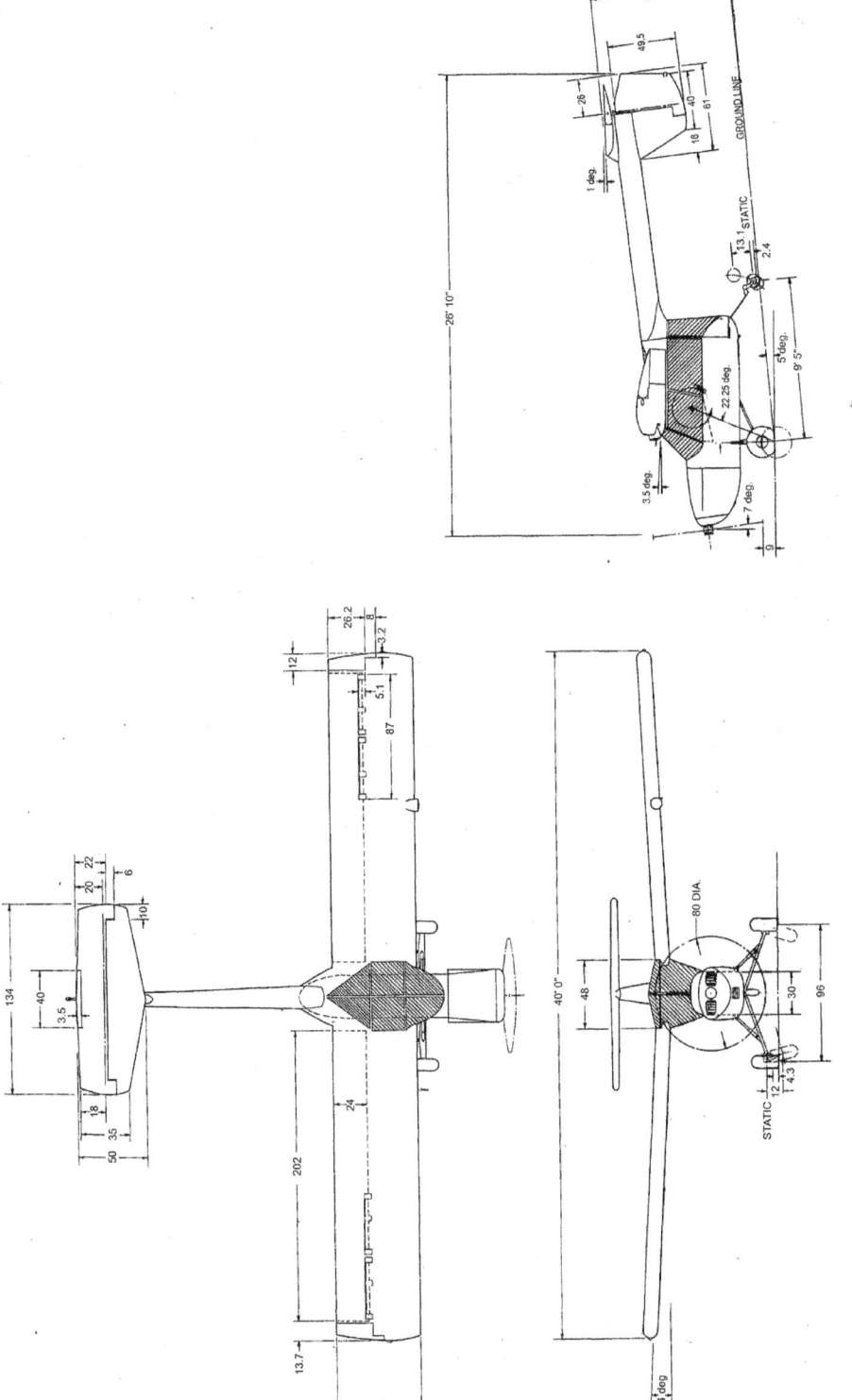

THREE-VIEW LINE DRAWING
FIGURE NO. 1

Boeing Model 451-2 three-view (author's collection).

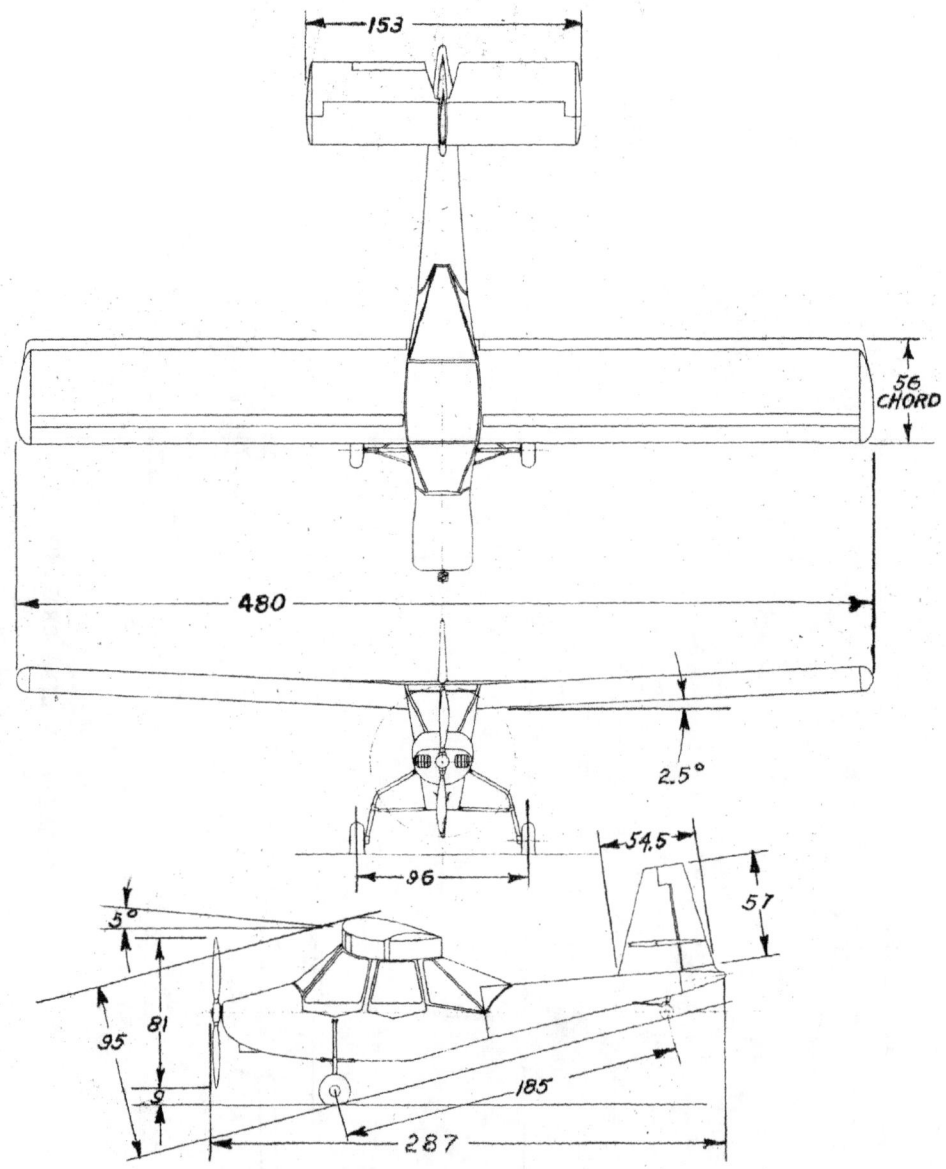

Consolidated Vultee proposal three-view (NARA).

The nose gear was steerable with 5.00 × 4 wheels. Provision was made for the installation of skis, pontoons, or Brodie landing gear.[5]

Boeing (Wichita Division)

During World War II, Boeing Wichita had built thousands of PT-17 biplane single-engine primary trainers and B-29 four-engine bombers. They had no liaison aircraft experience, but as they were out of work and income due to end-of-war cancelations, they had been doing a light plane design study. The Boeing proposal to the design competition was the Model 451–2 and based on their Model 451 personal aircraft preliminary design.

It appears that Boeing had an advantage over the other competitors. In an informal conference with ATSC Engineering Division personnel in early September 1945, they found out about the upcoming design competition and began their preliminary design study for the aircraft. On December 5, 1945, Boeing submitted their proposed design to ATSC for comments.[6]

Because of their advanced notice of the competition, Boeing's submitted proposal documentation had in the order of ten times as many pages compared to those of the other submitters who had only a month to do their proposals.

The proposed Model 451–2 officially submitted to the competition in March 1946 was a high wing single-engine tractor all-metal two-place aircraft with tandem seating. The fuselage was unusual in that it had a gondola-like cabin with a door on the right side for the pilot while the observer's fully transparent door was on the aft end of the gondola. A small diameter tail boom went slightly upwards from the gondola and supported a horizontal stabilizer and elevator with a single inverted vertical fin and rudder below.

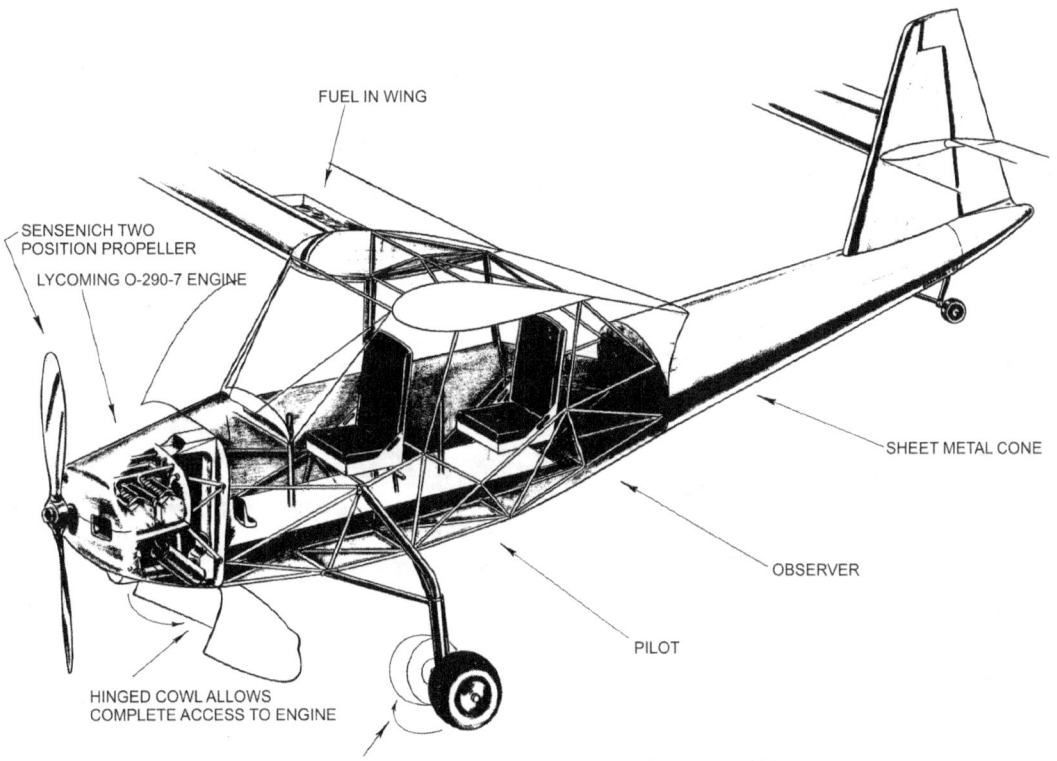

Consolidated Vultee proposal cutaway (NARA).

The proposed Model 451–2 had a rectangular cantilever wing, with large 40 percent chord Fowler flaps instead of the external flaps used on the Model 451 personal plane design. Lateral control was by circular arc scoop–type spoilers and feeler ailerons that provided stick force.

The landing gear was a fixed conventional gear with a tail wheel. The main gear consisted of a welded steel vee with oleo shock absorbers. Due to the unusual gondola fuselage, the tail wheel was not mounted below the tail as usual but mounted below the aft end of the gondola. There were provisions for floats and skis, and in addition, Brodie gear could be fitted.

The Boeing proposal was powered by a Lycoming O-290–7 engine with an 80-inch diameter Sensenich two-position controllable pitch propeller.[7]

An in-depth description and discussion of Boeing's Model 451–2 entry is given in the following chapter.

Consolidated Vultee (San Diego Division)

The Consolidated Vultee Aircraft Corporation (Convair) was experienced in designing and producing liaison aircraft, but that experience was all in their Stinson

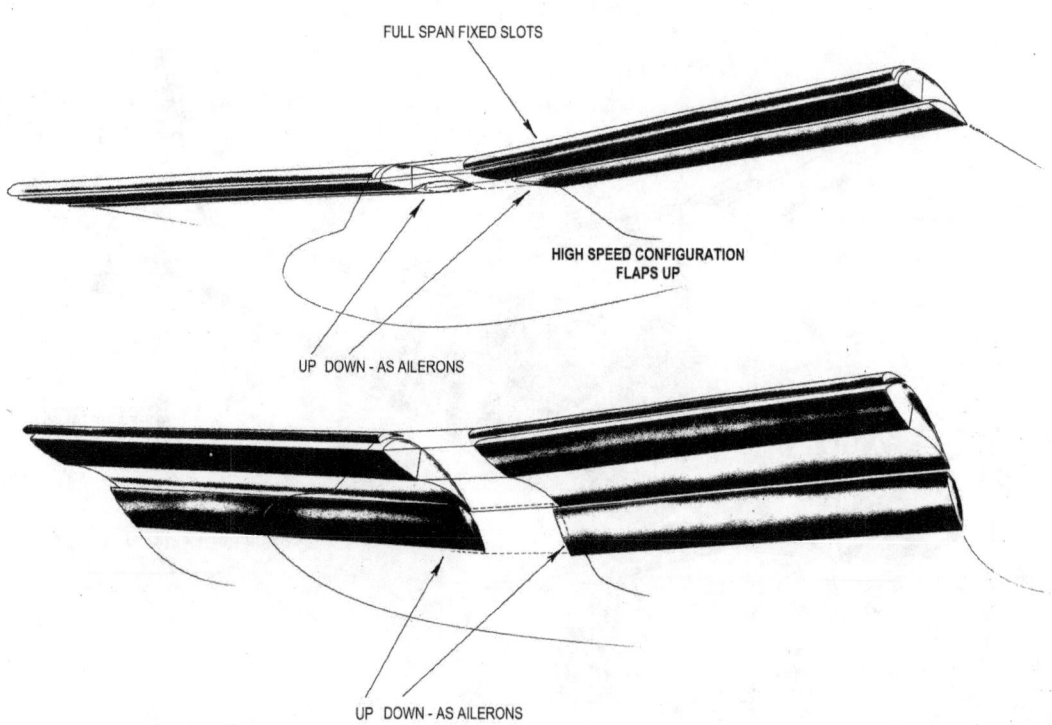

Consolidated Vultee proposal wing and flap (NARA).

division in Wayne, Michigan. Surprisingly, the headquarters San Diego division, best known for multi-engine bombers, sent in the bids for this light airplane. Also, Convair's (Stinson's) liaison aircraft experience was primarily with the Army Air Forces, not the Army Ground Forces.

Consolidated Vultee submitted two bids, but they were for the same airframe with the same basic engine, one being the standard Lycoming O-290–7 with a carburetor and the other being a version of the engine under development with fuel injection. The proposed aircraft had a configuration that looked somewhat like a

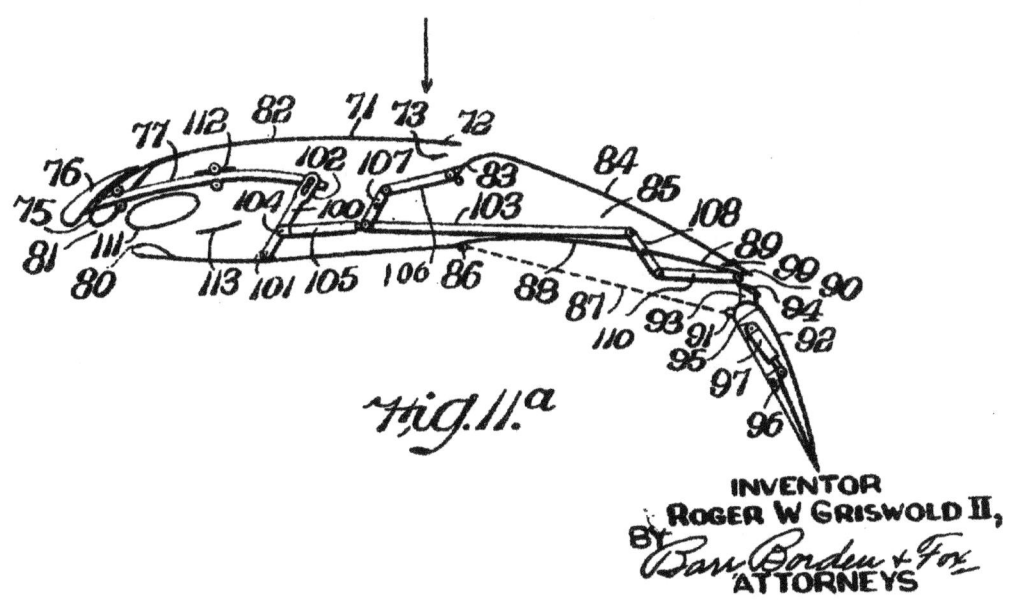

Top: Ludington-Griswold XLG-3 airplane with its patented Integrated Wing (Ted Koch).
Bottom: Roger Griswold's Integrated Wing (U.S. Patent 2,348,253).

shrunk-down XL-13, a liaison/ambulance airplane that their Stinson division was developing for the Army Air Forces. The tail configuration was copied from the XL-13, except it had no strut bracing.

The wing was a different configuration from the XL-13 and was a rectangular platform and fully cantilevered with no struts. The wing had a full-span fixed leading edge slot and a full-span 50 percent chord Fowler flap. The upper wing skin went back to 90 percent chord. There were no ailerons. Lateral control was by using the Fowler flap as a flaperon. In the retracted or flap up position, the flaperon could deflect down but not up. In the extended or flap down position, the flaperon could deflect up or down.

The landing gear was a conventional tail wheel configuration. The main wheels rotated 180° for truck loading. There was provision for the installation of skis, pontoons, or Brodie landing gear.[8]

Ludington-Griswold, Inc.

Charles Townsend Ludington and Roger W. Griswold II were aeronautical engineers who had worked together on several occasions. They are best known for designing the Pratt-Read PR-G1 glider (Navy LNE-1, Army TG-32). The two of them formed Ludington-Griswold, Inc., in Saybrook, Connecticut, for aeronautical research and aircraft development. Ludington-Griswold built two research aircraft. One was a three-quarter-scale test aircraft (registered NX60333) of a four-place mid-wing pusher amphibian which flew only once. The other was the XLG-3, a modified Fairchild 22 (registered NX14768) with an experimental LG Integrated Wing with full-span multi-flaps and endplate fins with control surfaces.

Ludington-Griswold submitted their Model XLG-31 to the design competition. At first glance, their proposal looked like a conventional strut-braced high wing two-place tandem seat light plane, but it was actually a very complex design.

The major complexity was the Ludington-Griswold Integrated Wing that used a GS-13 airfoil section with full-span slat and flap and partial-span outboard spoiler panels. The unique feature of the Integrated Wing is that the leading edge has an inlet that allows for air to flow internally in the wing and could be directed to go up the slot of the slat, the flap, and or spoiler. The flap system was spring-loaded to automatically give flap deflections appropriate to speeds in the lower speed range. This would significantly reduce ship pitching angles over a wide range of speeds, permitting a perfectly normal flare-out technique with a highly flapped airplane and enabling the ship to accelerate on the optimum flight path to the speed for best climb. An overriding manual control was provided to lock the flaps in the desired position to give the pilot a selective range of landing speeds. Lateral control was by new-type spoilers developed by Ludington-Griswold, which control the internal flow through the Integrated Wing. The Integrated Wing's excellent characteristics had previously been demonstrated on the Ludington-Griswold XLG-3 airplane, especially the highly effective lateral control at low speeds and full flap deflection. These spoilers could be arranged to operate together to control the glide path and

3. XL-15 Liaison Aircraft Competition

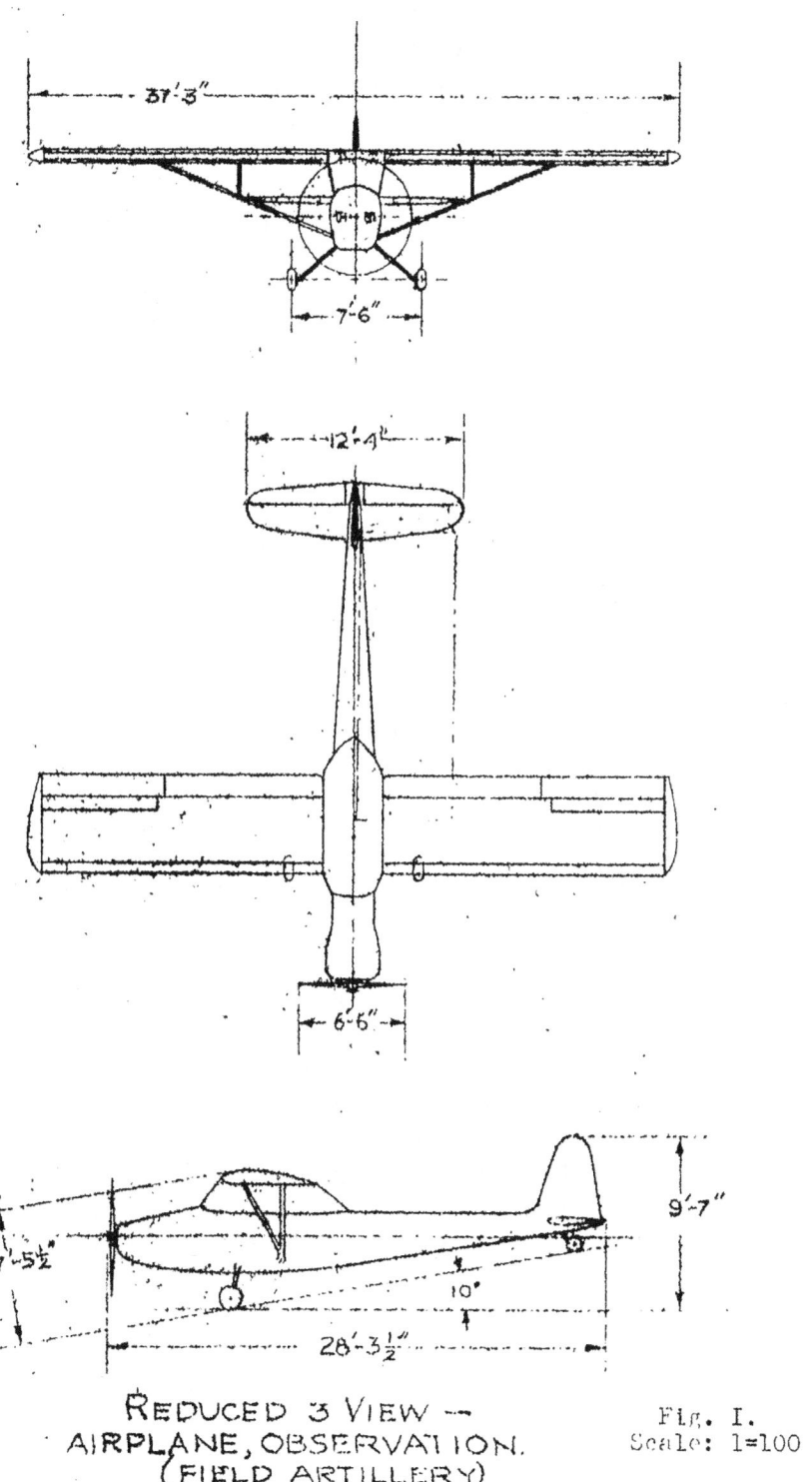

Ludington-Griswold Model XLG-31 three-view (NARA).

simultaneously give opposite action for lateral control. The features of the Integrated Wing were covered by Roger Griswold's U.S. Patent 2,348,253.

The horizontal tail was of the all-movable type and provided longitudinal control and stability plus properly coordinating the change of trim, power on and power off and at all flap deflection angles to simplify piloting technique. Dual flight controls were provided. The vertical tail shown on the three-view drawing does not show a rudder and may also be all-movable like that tested on a NACA Fairchild Model 22.

The LG-31 was to be spin-proof and impossible to stall out of a turn.

The airframe was all-metal construction and designed to CAA requirements. It had wings that folded back for easy storage. The landing gear was of conventional tail wheel type with oleo shock absorbers. There were provisions for the installation of skis, pontoons, and Brodie landing gear.

The power plant was a 140 hp fuel-injected version of a Lycoming GO-290 geared four-cylinder engine turning an 80-inch diameter two-position propeller. Prop rpm was 0.64 crankshaft rpm. The engine had a self-starter.[9]

Piper

Piper Aircraft Corporation of Lock Haven, Pennsylvania, was the world's most experienced company in developing and manufacturing liaison airplanes. During World War II, Piper had produced the Army Ground Forces' primary liaison aircraft, the L-4, and its chosen replacement, the L-14 (which had been canceled when the war ended).

Piper's entry, the PA-9, was a strut-braced constant chord high wing tandem seat aircraft of similar configuration to the L-4. However, when looked at closely, there were many differences.

The wing used the same chord and airfoil as used on the L-4 and L-14. However, the wing tips were of a squared-off configuration using a detachable wing tip formed from an aluminum sheet instead of the circular bowed tips used on previous Piper airplanes. The wing structure consisted of two 61ST extruded aluminum spars with wing ribs fabricated from drawn aluminum sections riveted into a truss-type frame and attached to the metal spars by self-tapping screws. The forward rib leading edges were covered with sheet aluminum, while the wing panels were fabric covered. The wing had slotted flaps and ailerons that could be drooped 20°. The PA-9 wing was designed for quick and easy removal with quick disconnects for controls, wires, and fuel lines.

Both fixed and movable tail surfaces were constructed of steel tubes with ribs fabricated from steel channels and were fabric covered. The PA-9 used a controllable sheet metal elevator tab for trim in place of the adjustable incidence stabilizer used on the L-4.

The fuselage structure was a welded steel tubular framework covered with fabric. The rear fuselage was cut back more than on the L-4 and provided better visibility.

Top: Piper L-4B (author's collection). *Bottom:* Prototype Piper YL-14 that was canceled on the promise the L-15 would be an improvement (NARA).

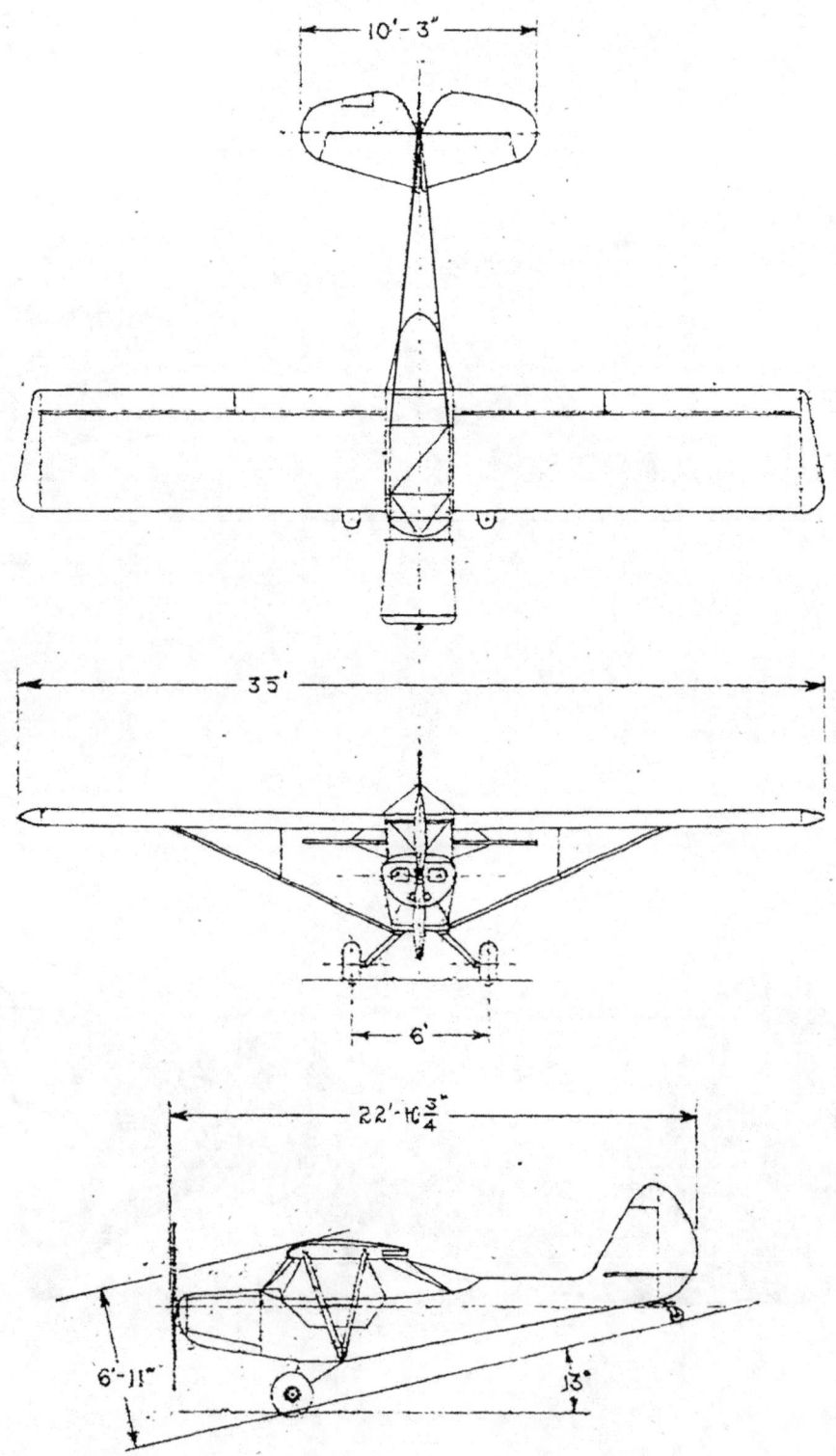

Piper PA-9 three-view (NARA).

The landing gear was similar in appearance to that used on the L-4 with two significant differences: larger tires (8.5 × 6) and hydraulic stock absorbers instead of bungees. There were provisions for floats, skis, and a Brodie landing system as on previous Piper liaison airplanes.

Piper differed from other competitors by not offering the preferred 125 hp Lycoming O-290 engine. Piper had used the Lycoming O-290 on their L-14 liaison airplane. Based on that experience, Piper engineering felt that the O-290 was too low in power to meet the competition's field performance and climb requirements. They offered their entry (PA-9–165) with the Continental E-165 engine (165 hp Normal Rating and 185 hp Take Off Rating) and also with an alternate engine (PA-9–190), the Lycoming O-435–11 engine (190 hp Normal and Take Off). Piper used the rated powers for takeoff and climb, but they used 94 hp instead of 75 percent of the rated engine power for cruise speed, range, and endurance. Although 94 hp is 57 percent power on the E-165 and 49 percent power on the O-435–11, it was 75 percent power for the O-290. By doing this little trick, they had more than enough power for takeoff and climb while still having enough power along with low fuel flow to meet the cruise, range, and endurance requirements.

Both engine offerings used an 84-inch diameter Beech-Roby constant-speed two-bladed propeller.[10]

Ratings

The following is a reprint of the ratings and discussion of the completion entries transmitted by the Air Materiel Command (the new name of the Air Technical Service Command) to Carl Spaatz, Commanding General, Army Air Forces, on April 18, 1946.

DISCUSSION OF ENGINEERING AND SUITABILITY CHARACTERISTICS—MANUFACTURER'S RATINGS

No. 1—PIPER (PA-9–190)—By virtue of its high strength-weight ratio, good spin characteristics, and excellent adaptability to quantity production, this design rates next to highest in Structure and Design. However, Piper's choice of the 190 horsepower engine, as well as the lack of provisions for All-Weather Operations, accounts for the next to the lowest award for Power Plant Installation. (All manufacturers were advised that a 125 horsepower engine was desired by the Army Ground Forces for reasons of simplified field maintenance, and that the choice of a larger engine would correspondingly detract from the proposal's merit. They were also advised that provisions should be made for satisfactory operation of the engine in extreme cold, tropical, dusty, or salty atmospheres.) Maintenance and Repair features are satisfactory but Equipment Installation is considered only average. Suitability Characteristics are not considered desirable by reason of its average visibility and operation from small, unprepared fields—the latter being adversely affected by its high landing speed. In as much as Piper's developments have been along commercial rather than military lines and since their financial status is considered healthy, they have been given two-thirds of M_{max} for Background and financial status. Their commercial backlog being quite large, only forty percent of M_{max} is awarded on Work Load.

No. 3—BOEING-WICHITA (451)—This design offers the maximum in the desired configuration and arrangement. By virtue of Boeing's experience with Army design requirements and production of PT-17's, very satisfactory results may be expected insofar as design criteria,

Table 3.3. Proposal Data Comparison

Company			Bellanca	Bellanca	Boeing	Convair	L-G	Piper	Piper
Model No.			20-13	24-19	451-2	A & B	XLG-31	PA-9-165	PA-9-190
Wing	Area	sq ft	200	240	200	186.67	216	180	180
	Span	inches	436	465	480	480	447	420	420
	Chord	inches	66	74.2	60	56	72	63	63
	Aspect Ratio		6.6	6.3	8	8.57	6.3	6.8	6.8
	Airfoil		23015	23015	4415		GS-13	USA35B	USA35B
	Root Incidence	°	4	4	3.5	5	0	0	0
	Dihedral	°	2	2	4	2.5	0	0.75	0.75
	LE Slat		fixed	fixed	none	fixed	movable	none	none
Ailerons	Area	sq ft	15	18	3.4	Flaperon	39	18.76	18.76
	Deflection up/dn	°	30/15	24/19	25/25			30/20	30/20
Spoiler	Type		none	none	Circular Arc	none	Int.Flow	none	none
	Span (each)	inches			87				
	Deflection	°			60				
Flap	Type		Dbl Slot Flap Drooped Ail	Dbl Slot Flap Drooped Ail	Fowler	Full-span Fowler	Full-span Dbl Sloted	Slotted Flap Drooped Ail	Slotted Flap Drooped Ail
	Deflection main/aux	°	40/68	40/68	40			45	45
	Main Flap Chord	% mac	25.66	25.66	40	50		21	21
	Auxiliary Flap Chord	% mac	10	10					
Horizontal Tail	Area	sq ft	40	48	40		36	30.16	30.16
	Span	inches	171	186	134	153	148	123	123
	MAC	inches	35	38	43				
Stabilator	Area	sq ft	24.6	29.4	22.7			15.76	15.76
	Incidence	°	-3	-3	-1			0	0
Elevator	Area	sq ft	15.4	18.6	17.3			14.4	14.4
	Deflection up/dn	°	30/20	30/20				30/25	30/25
Vertical Fin	Area	sq ft	7	8.4	7.3		13.5	6.05	6.05

Company			Bellanca	Bellanca	Boeing	Convair	L-G	Piper	Piper
Rudder	Area	sq ft	7	8.6	7.2			7.5	7.5
	Deflection R/L	°	27.5/27.5	27.5/27.5				20/20	20/20
	Tab		fixed	fixed	fixed			controllable	controllable
Fuselage	Max Height	inches	55	55	48		40	57.75	57.75
	Max Width	inches	34	34				34	34
	Aircraft Length	inches	299	320.5	322	287	321.6	274.75	274.75
	Aircraft Height	inches	103	110	102	95	98	83	83
Power Plant	Manufacturer		Lycoming	Lycoming	Lycoming	Lycoming	Lycoming	Continental	Lycoming
	Model		O-290-3	O-435-11	O-290-7	O-290-7	GO-290	E-165	O-435-11
	Normal Rating	BHP/rpm	125/2600	190/2550	125/2600	125/2600	140	165/2050	190/2550
	TO Rating	BHP/rpm			130/2800	130/2800		185/2300	
Propeller	Type		controllable	controllable	2 position	2 position	2 position	const speed	const speed
	Material		wood	wood	wood			wood	wood
	Diameter	inches	80	84	80	81	80	84	84
Weight	Design Gross	lbs	1,926	2,234	2,050	1,900	1,815	2,000	2,030
	Alternate Gross	lbs	2,020	2,328		1,998	1,915	2,112	2,142
	Empty	lbs	1,299	1,531	1,509	1,305	1,226	1,353	1,383
Performance	Cruise Speed	mph	95	100	109	90	105	104	104
	Endurance with res	hrs	3.5	4	4	3.5	3.75	4.67	4.47
	Service Ceiling	ft	15,500	18,000	17,400	16,000	18,700	19,300	20,000
	max Rate of Climb	fpm	675	800	690	675	1,080	1,130	1,170
	Landing Speed	mph	31.5	30	35	35	31	42.2	42.5
	TO over 50 ft, sod	ft	470	300	595	575	386	550	525
	Landing over 50 ft	ft	325	400	517	432	335	550	550
	Ferry Range, ext tank	miles	500	500	560	500	500	740	710

simplicity of design, spin characteristics, safety features, and adaptability to quantity production are concerned. The Power Plant Installation has been awarded the largest number of points, principally on the Boeing Company's choice of an approved 125 horsepower engine. Fuel consumption for this airplane is considered the least of all the designs submitted. The airplane's configuration, production breakdown, and quick disconnect features give promise of an exceptionally fine airplane from a Maintenance and Repair standpoint. Past experience with the Boeing Company engineers has proved their capabilities in the radio, instrument, electrical, and miscellaneous equipment installation field, so that they have been awarded the highest rating for Equipment installation. Operation from small, unprepared fields is considered above average due to its small size, weight, and landing gear arrangement. The visibility of both the pilot and the observer is substantially better than any other design presented. Having provisions for All-Weather Operations and quick disconnect features, and excellent accessibility to cockpits, makes this design entirely suitable for its mission. Due to the Wichita Division's position with regard to backlog, they have received the maximum Figure of Merit for Workload, which gives them a slightly higher Manufacturer's Rating than the Stinson Division of Consolidated-Vultee.

No. 4—BELLANCA (20-13)—Structure and Design features are satisfactory. However, complexity of design and operation makes this aircraft only average engineering-wise as compared with the other proposals submitted. It has the largest fuel consumption of any design submitted and has no All-Weather and self-sealing provisions. Maintenance and Repair items offer serious difficulties. Due to Bellanca's inexperience with the equipment required for this Army Ground Force airplane, any optimism regarding the installation of such equipment would be unwarranted. Performance-wise the airplane seems to be underpowered and due to its tricycle landing gear its operation from small, unprepared airfields is questionable. Furthermore, due to the airplane's conventional configuration, visibility of both the pilot and the observer is impaired. This company's commercial and military backlog of work is not large. Its financial status is considered healthy. Background for this type of aircraft is weak.

No. 5—BELLANCA (24-19)—This proposal employs a 190 horsepower engine at a penalty. It promises to give the best performance of all proposals submitted. However, its suitability is considered the lowest due principally to its large size, weight, and landing gear arrangement. Otherwise, it ranks the same as No. 4.

No. 6—CONSOLIDATED-VULTEE (A)—This design offers about average efficiency from a strength-weight ratio standpoint. However, its conventional configuration does not lend itself particularly to the type of mission required. The design closely follows that of the Army Air Forces Liaison Type XL-13. The fuel injection engine employed has little chance to be approved prior to installation in an aircraft. Fuel consumption is high. Maintenance and Repair features as well as Equipment Installations are good but are considered less satisfactory than the Boeing proposal. Operation from small, unprepared fields is good due to its low landing speed. However, visibility is considered barely satisfactory. Its adaptability to alternate uses does not approach that of the Boeing design. Past experience with this manufacturer and consideration of his financial status justifies a high rating for Background and Financial Status. However, Consolidated-Vultee's Work Load on military contracts alone is heavy, consequently, it is believed to be in the Government's interests not to award this contract to this manufacturer.

No. 7—CONSOLIDATED-VULTEE (B)—This proposal is essentially the same as No. 6, except that a standard carburetor is used in lieu of fuel injection for which full credit for choice of power plant has been awarded.

No. 8—LUDINGTON-GRISWOLD–(XLG-31)—This proposal ranks next to the lowest on a strength-weight ratio basis. The manufacturer's unfamiliarity with the Handbook of Instructions for Airplane Designers and other Army Air Forces design requirements disqualifies him to design satisfactory Power Plant Installations, All-Weather provisions, self-sealing

Piper PA-9 inboard profile (NARA).

tanks and lines, radio, instrument, and electrical installations, armor provisions, etc. His inexperience in the design of complete aircraft adversely affects the simplicity of design, adaptability for production, ease of inspection, Maintenance and Repair, rapid assembly and disassembly features. The configuration of the design submitted is conventional and offers no distinct advantage insofar as the mission to be performed is concerned. The principle of the integrated wing is untried—and falls into the category of a research project. Fuel economy is low. Visibility of the pilot and the observer is only fair. Background and Financial status for development of Army Air Forces equipment is considered poor.[11]

Chosen Winner

Although the Piper PA-9 had the highest rating, it was not the chosen winner of the Informal Design Competition of Field Artillery Observation Aircraft. This is not an unusual occurrence for those of us who have spent much time in the aircraft industry.

The following excerpt is the recommendation transmitted by the Air Materiel Command to Carl Spaatz, the Commanding General, Army Air Forces on April 18, 1946.

> 3. It is the opinion of the Air Materiel Command that the Boeing Aircraft Company, Wichita Division, will produce, around their design of the Model 451, the aircraft which will most nearly fulfill the requirements under which this aircraft is being procured. Consequently, it is the recommendation of the Air Materiel Command that a contract be entered into between the Government and the Boeing Airplane Company, utilizing 1946 Research and Development Funds in the amount of $325,597, to procure one (1) basic flight test article, one (1) complete aircraft, and such other items as are considered essential to the development of a successful experimental aircraft.
>
> 4. Since it is urgent that funds be obligated out of the current fiscal year's program, it is requested that the Air Materiel Command be notified of your decision at the earliest possible date.[12]

The Air Materiel Command–recommended contract amount matched Boeing's bid, which is broken down in Table 3.4.[13]

Table 3.4. Boeing Bid

Item No.	Item	Cost
1	Flight Articles (2 airplanes)	$93,390
2	Mockup	$9,000
3	Data	$193,582
4	Handling Aids	
	a. Handling Aids	$500
	b. Handbooks	$6,800
5	Tooling	$12,125
6	Flight Tests	$10,200
	Total Bid	**$325,597**

Headquarters Army Air Forces indicated that Air Materiel Command could take action in awarding a contract to Boeing utilizing 1946 Research and Development Funds, but further stated that the original informal price quotation appeared excessive and directed Air Materiel Command to clearly analyze the price quotation prior to issuing the final contract.[14]

A letter contract was issued to Boeing on July 22, 1946. The definitive contract W33–038-ac-15054, dated February 27, 1947, for two XL-15 aircraft, No. 1 stripped to be used after flight test for a static test, and No. 2 a completely equipped airplane, one mockup, and the normal data and spare parts, at an initial cost of $325,597. It must be noted that many items such as engines, propellers, radios, wheels, and tires were government furnished equipment (GFE) and not included in the above cost of the contract. The total cost of the GFE for the two XL-15 prototypes was $12,106.23.[15]

4

Boeing's Winning Entry

In an informal conference with ATSC Engineering Division personnel in early September 1945, Boeing had found out about the upcoming liaison aircraft design competition and began a preliminary design study for such an aircraft. On December 5, 1945, Boeing submitted their proposed design to ATSC for comments.[1]

Boeing's 451-2 proposal was finalized in February and March 1946. The short time period of the L-15 liaison aircraft Informal Design Competition, which only allowed bidders 30 days to submit their proposed design, gave large military-oriented companies like Boeing an advantage. They were used to the way Wright Field wanted things documented. Boeing also had the benefit of advanced information and discussion with Wright Field on the competition. Boeing received the official type specification on February 6, 1946. Earl Weining was the Boeing Project Engineer for the program from beginning to end.

Boeing's entry, the Boeing Model 451-2, was presented in several company documents, including WD-12057 Detail Model Specification, Boeing Model 451-2, and WD-12056 Preliminary Performance, Stability and Control Characteristics, dated March 6, 1946. WD-12056 was 202 pages thick and illustrated the capabilities and manpower that a company like Boeing could use to respond to a request for proposals. Boeing also submitted a weight report, a structures report, and design drawings for the proposed airplane, its structure, equipment, and systems. Boeing finished up its proposal on March 6, 1946, in time to meet the March 8, 1946, deadline. During the month, while preparing its bid, Boeing expended 4,902 man-hours.

The Wichita division's look at a light plane was the beginning basis of their design proposal. However, no consideration was given to the two-control system as this feature was considered to be in too elementary a design stage at the time. Also, external flaps were rejected and replaced with Fowler flaps as Boeing was very experienced with Fowler flaps. But the canted thrust line to minimize power effects and the high horizontal tail above the wing's wake that had been developed using the L-6 testbed were both incorporated.

Like most military aircraft proposals, Boeing's entry was different in many ways from the aircraft that would later fly as the Boeing XL-15. The proposed Model 451-2 was a high wing single-engine tractor all-metal two-place aircraft with tandem seating with a gondola-like fuselage. The landing gear was of a conventional layout with

Boeing Model 451–2 (*The Field Artillery Journal*).

a tail wheel. The empennage consisted of a horizontal tail and an inverted vertical tail mounted on the end of a boom that extended aft of the cabin.[2]

Wing

The smooth, all-metal cantilever wing was rectangular in planform and a high wing configuration mounted on top of the fuselage. A landing light was mounted in a headlight-like fairing on the leading edge of the left wing, and navigation lights were mounted on each wingtip. The wing had Fowler flaps large in both chord and span to provide very high lift (a landing C_{Lmax} of 3.27). The flaps had four positions: (1) Retracted—no Fowler motion and no deflection, (2) Slow Flight–Fowler motion (flap nose at 97 percent wing chord) and 0° deflection, (3) Take Off–Fowler motion and 25° deflection, and (4) Landing–Fowler motion and 40° deflection.

Because of the large span flaps, the Boeing Model 451–2 did not have conventional ailerons. For lateral control, it used spoilers and feeler ailerons. The spoilers were of a circular arc scoop style that rotated into the airstream above the outboard section of the wing. The ailerons were of a large chord, but with a very small span at the extreme outboard end of the wings, and provided lateral stick force. The spoilers provided 72 percent of the max roll rate and the ailerons 28 percent. Of the lateral stick force the pilot felt, the spoilers provided only 1 percent, and the ailerons provided 99 percent.[3]

Table 4.1. Wing Geometry

Span	480 inches
Mean Aerodynamic Chord (actual)	60 inches
Area (actual)	200 sq ft
Airfoil	NACA 4415
Leading Edge of MAC	FS 81.78, WL 138.46
Incidence	3°30"
Dihedral	4°
Aspect Ratio (actual)	8.0
Taper Ratio	1.0

Table 4.2. Flap

Type	Fowler
Span (each)	202 inches
Chord	24 inches (40% c_w)
Area	67.33 sq ft
Inboard End	BL 24.3 (10 % $b_w/2$)
Outboard End	BL 226.3 (94 % $b_w/2$)
Deflection (Slow Flight)	0°
Deflection (Take Off)	25°
Deflection (Landing)	40°

Table 4.3. Aileron

Type	plain sealed with horn balance
Span (each)	12 inches
Chord	26.2 inches (43.8% c_w)
Area aft of hinge line (each)	1.7 sq ft
Horn Area ahead of hinge line (each)	0.2 sq ft
Inboard End	BL 228 (95 % $b_w/2$)
Outboard End	BL 240 (100 % $b_w/2$)
Deflection	± 25°
Tab Type	none

Table 4.4. Spoiler

Type	circular arc scoop
Span (each)	87 inches
Projected Scoop Area (each)	5.2 sq ft
Chordwise Location	66% c_w
Hingepoint Location	56.4% c_w
Inboard End	60 % $b_w/2$
Outboard End	95 % $b_w/2$
Deflection	60°
Maximum scoop extension	4.62 inches

Horizontal Tail

The all-metal horizontal tail had a tapered planform with a one-piece elevator. The stabilizer was fixed, and the elevator had a controllable tab to provide longitudinal trim. The elevator and elevator trim tab hinge line was straight with zero sweep angle. The elevator leading edge was circular, with aerodynamic and weight balance being provided by horn balances at the tips. The horizontal tail used an inverted Clark Y airfoil and was mounted on top of the tail boom.[4]

Table 4.5. Horizontal Tail Geometry

Span	134 inches
Root Chord	50 inches
Tip Chord	35 inches
Mean Aerodyamic Chord	43 inches
Area	40 sq ft
Airfoil	inverted Clark Y
Incidence	-1°
Dihedral	0°
Aspect Ratio	3.12
Taper Ratio	0.7
Tail Arm at Design C.G. (32% M.A.C.)	16.7 feet

Table 4.6. Elevator

Type	plain with horn balance
Span	134 inches
Root Chord	20 inches
Tip Chord	18 inches
Area aft of hinge line	17.3 sq ft
Horn Area ahead of hinge line	1.1 sq ft
Tab Type	plain
Tab Span	40 inches
Tab Chord	3.5 inches
Tab Area	0.97 sq ft

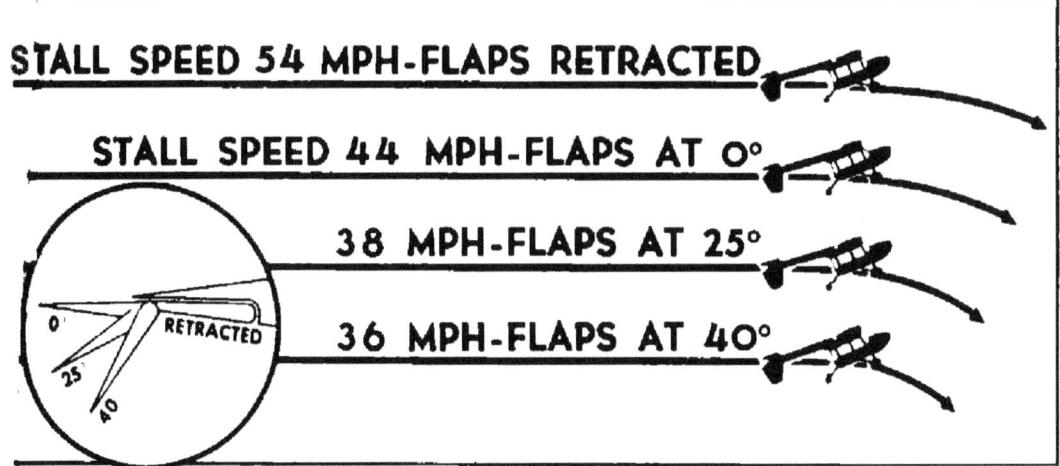

Boeing Model 451-2 stall speed (*The Field Artillery Journal*).

Vertical Tail

The all-metal vertical tail was unusual in that it was inverted and mounted below the tail boom. Otherwise, except for appearing upside down, it was a normal single vertical tail with a rudder that extended behind the tail boom and was cut away to provide clearance for a downward deflected elevator. The rudder leading edge was circular, with aerodynamic and weight balance provided by a horn balance at the tip. The rudder hinge line was perpendicular to the axis of the tail boom. The rudder had a fixed tab that could be adjusted for trimming. The rear navigation light was mounted on the rudder trailing edge.[5]

Table 4.7. Vertical Tail Geometry

Type	single inverted
Root Chord	61 inches
Tip Chord	40 inches
Mean Aerodyamic Chord	49 inches
Area	14.5 sq ft
Airfoil	BAC-66
Aspect Ratio	1.0
Taper Ratio	0.66
Tail Arm at Design C.G. (32% M.A.C.)	16.7 feet

Table 4.8. Rudder

Type	plain with horn balance
Span	49.5 inches
Root Chord	26 inches
Tip Chord	21 inches
Area aft of hinge line	7.2 sq ft
Horn Area ahead of hinge line	0.5 sq ft
Tab Type	fixed

Fuselage

The fuselage was like that of a typical high wing lightplane or liaison plane except that it bluntly ended at the rear of the cabin, and then a boom projected rearwards and upwards to which the empennage was attached. The fuselage was of all-metal construction. The crew was seated in tandem, with the rear observer's seat being able to face forward or rearward. There was extensive window glazing to provide 360° observation. The side windows were raked outward as they went up to give both crew members a good view down to the side. The wing center section was transparent to allow an overhead view. The cabin ended with a glassed vertical door that could be opened in flight for dropping things out. There was a provision on the lower fuselage to hook a tow cable so the aircraft could be towed like a glider.

The total length of the aircraft from the tip of the spinner to the rudder trailing edge was 26 feet 10 inches.[6]

Boeing Model 451–2 features (*The Field Artillery Journal*).

Power Plant

Boeing chose to use the Lycoming O-290–7 engine that was the recommended engine in the Request for Bids. The O-290–7 was a dual ignition four-cylinder, air-cooled, horizontally opposed aircraft engine with a 6.5:1 compression ratio.[7]

An 80-inch diameter Sensenich two-position hydraulically controllable pitch propeller using hub model number C-3FB1 was specified. The wooden blades had a Clark Y section with an activity factor of 121.6. It had a 7.5° low pitch setting and an 11.8° high pitch setting.[8]

Landing Gear

The landing gear in the proposal was a fixed conventional gear with a tail wheel. The Boeing Wichita division was the old Stearman Aircraft Corporation, and the main landing gear was very similar to that used on the early Stearmans (those before the Model 70 series aircraft). The main landing gear consisted of a welded steel landing gear vee that was pin mounted to the centerline of the lower fuselage, and oleo shock absorber struts that attached to the vee at the stub axle and went up to the side of the fuselage. The landing gear had a wheel thread of 96 inches and a wheelbase of 113 inches. The tail wheel was freewheeling and was attached to the rear of the fuselage gondola cabin by a long spring steel arm.[9]

In addition to the primary wheeled landing gear described above, there were provisions for floats and skis. In addition, Brodie gear could be fitted.

Structural Load Factors

Max positive load factor +5.0
Max negative load factor −2.5[10]

Performance

The performance estimates were done using standard methods and references of the time. The submitted preliminary performance report referenced each of the equations and data sources used.

Table 4.9. Guaranteed Performance Items[11]

Cruising speed at 75% power	109 mph
Endurance at operating speed (47% power), including ½ hour reserve fuel	4 hours
Service ceiling	17,400 feet
Rate of climb at sea level	690 fpm
Landing speed	35 mph
Take-off distance over 50-foot obstacle, operation from dry sod, short grass field	595 feet
Landing distance over 50-foot obstacle (design gross weight less one-half fuel)	517 feet
Ferry range with 24 gallons external tankage	560 miles

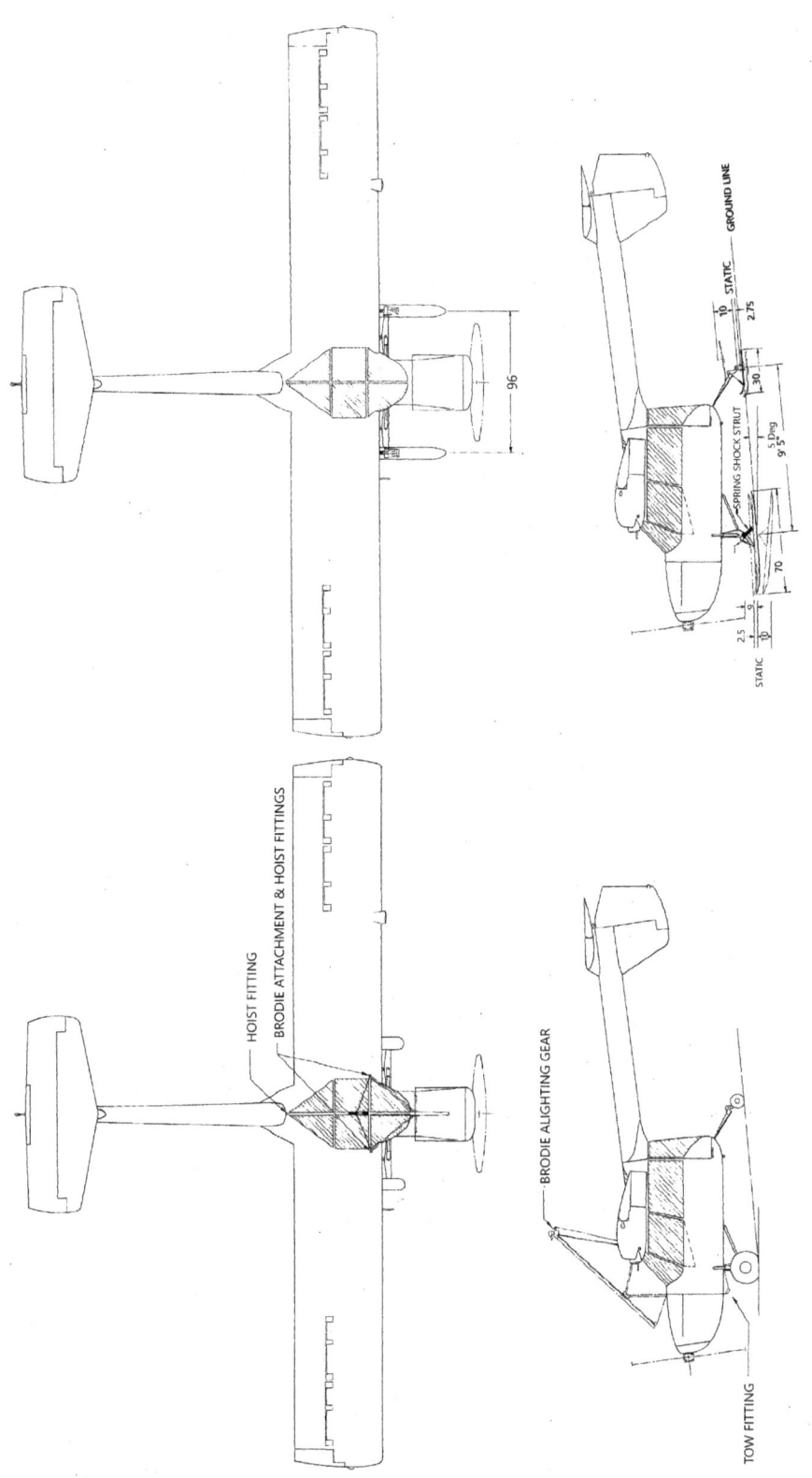

Boeing Model 451–2 with Brodie gear and skis (NARA).

Table 4.10. Additional Non-Guaranteed Performance Items[12]

Maximum speed at sea level	122.5 mph
Maximum speed at 5,000 feet	121 mph
Maximum speed it 10,000 feet	117.5 mph
Rate of climb at 5,000 feet	540 fpm
Rate of climb at 10,000 feet	355 fpm
Velocity in climb at sea level	72 mph
Velocity in climb at 5,000 feet	75 mph
Velocity in climb at 10,000 feet	78.5 mph
Time of climb to 5,000 feet	8 minutes
Time of climb to 10,000 feet	20 minutes
Absolute ceiling	21,400 feet
Stalling speed—25° flap setting	37.8 mph
Stalling speed—0° flap setting	44.2 mph
Stalling speed—retracted flap setting	53.8 mph
Endurance at cruising speed (75% Power)	5.25 hours
Operating speed (47% power)	89 mph
Cruising speed for ferry range (75% Power)	108 mph
Power-on landing distance over 50-foot obstacle	317 feet
Jet-assist takeoff distance over 50-foot obstacle	325 feet
Sinking speed at 1.10 Vs, power off	8.64 fps
Sinking speed at 1.10 Vs, power on	12.4 fps
Minimum speed of flight, extended flaps	57 mph
Terminal velocity at sea level (11.8° blade angle)	260 mph
Diving velocity at engine rpm of 3120 (11.8° blade angle)	183 mph
Maximum speed for being towed	165 mph IAS

Stability and Control

The aircraft's design center of gravity range was 25.4 percent MAC to 32.9 percent MAC. The estimated fixed stick neutral point was 48.8 percent MAC with flaps up and shifted 7 percent forward for flaps down, power on.[13]

The elevator stick force per velocity was estimated to be 1 lb. per mph when trimmed at 60 mph.[14]

Boeing used the methods given in various NACA reports to estimate and judge the longitudinal, directional, and lateral stability and control.

Weight

Performance analysis of the proposed aircraft design showed that the maximum weight could be no more than 2,050 pounds if the minimum performance requirements were to be met. A great percentage of the weight was set by the items specified in the Request for Bid.

Table 4.11. Items Established by Specification Requirements

Item	Weight, lbs
Electrical Equipment	102.6
Instrument Installation	31.2
Furnishings Installation	61.0
Starting System	22.6
Communicating Equipment	54.1
Crew	400.0
Propeller (fixed pitch metal)	35.0
Engine	258.8
Wheels and Brakes	60.5
Fuel (Minimum for 3-½ hours at operating speed)	120.0
Total	1145.8

Only 82 of the 1,146 pounds listed in Table 4.11 were contractor-supplied equipment or parts over which Boeing had some limited control. The result was that only 904 pounds were left for the airframe weight.[15]

5

XL-15 Testing and Design Refinement

The XL-15 became an official Boeing project in June 1946 when Boeing found out they had won the competition and would soon be awarded a contract to build two prototypes with AAF serial numbers 46–520 and 46–521 (Boeing construction numbers 20001 and 20002). XL-15 46–520 was used for aerodynamic and power plant development, while 46–521 was the fully equipped military standard prototype.

But before this date, Boeing Wichita had started some wing structural development testing that could apply to either a personal-type aircraft (Model 451) or the L-15 (Model 451-2). Nine different test panels were constructed to determine the shear strength of different types of chordwise beaded wing skins. These test panels had various numbers and heights of beads, two different skin thicknesses, different numbers of stringers, and different numbers of ribs. The analysis of the testing resulted in the decision to use conventional metal wing construction with unbeaded skins.[1]

At the time of contract award, the Air Materiel Command informed Boeing that the main landing gear needed to be redesigned to allow the loading of the airplane into a standard two and one-half-ton truck without the removal of the landing gear. The fix was to redesign the main gear to permit rotation of the wheels about a vertical axis through 180° like that of the Consolidated Vultee proposal. The maximum width of the main landing gear with the wheels rotated inward would not exceed 79 inches. The redesign resulted in a cantilever gear design with the shock absorber pistons acting as swivels. The redesigned XL-15 landing gear permitted a change in tread from 87.70 to 61.70 inches or some 26 inches.[2]

Following the contract award, Boeing Wichita sent its Chief of Flight Test, Elton Rowley, to Fort Sill to study the Army's tactical operation of liaison aircraft. Also, a tech rep from Boeing's Service Department visited various Army Ground Forces bases to see how liaison airplanes were operated and maintained by the Army. The tech rep also went to Alaska for the winter maneuvers to study the operation of liaison aircraft in cold weather conditions.[3]

Wind Tunnel Testing

Once Boeing learned that they were the winner, they constructed a 1/12.5 scale wind tunnel model to be wind tunnel tested in the University of Wichita's four-foot-diameter wind tunnel during the next four months. The wind tunnel tests showed there was a dead air spot aft of the cabin. This dead air resulted in an area of neutral stability approximately 6° to either side of the zero yaw condition resulting in no air over the vertical tail and rudder. The fix was to change from the single vertical tail to a twin tail. Because the twin tail end plated the horizontal tail, making it more effective, the horizontal tail was reduced in size, resulting in the new tail configuration being implemented with little weight increase. The twin tail arrangement has a disadvantage in that the rudders are out of the propeller slipstream, but the configuration improved rearward vision.[4]

The change to the twin inverted vertical tail configuration was also expected to be beneficial in spin recovery. The fin and rudders were located below the horizontal tail and would be completely free of any blanketing effect from the horizontal tail.[5]

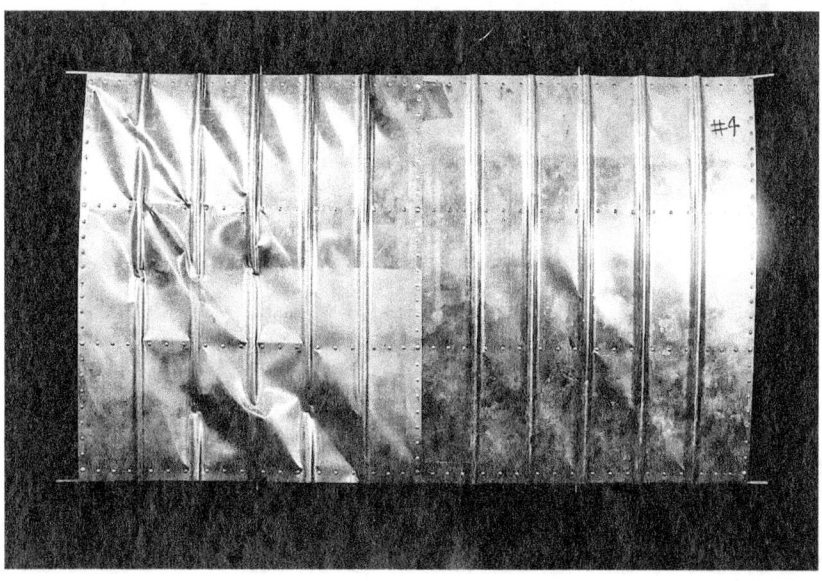

Top: Structural tests were run on partial wing panels of several different designs (author's collection). *Bottom:* Failure of a partial wing test panel (author's collection).

An inverted "V" tail arrangement was also looked at in the wind tunnel and found to be better than the twin tail from a stability standpoint. The downside of the "V" tail was that the control system was much more complex than that of the twin tail and would require more development both mechanically and aerodynamically. The "V" tail would have required additional wind tunnel, ground system, and flight testing. The complications, time, and cost needed for the "V" tail development were too much for the XL-15 program. However, it was recommended that the inverted "V" tail arrangement be given serious consideration in future aircraft of this type.[6]

Also, the wind tunnel tests showed that Boeing's initial Fowler flap wing with spoiler and feeler ailerons as specified in their winning bid did not work. They redesigned the wing based on the external airfoil flap wing of their previous personal airplane study. They kept the same primary wing planform with a 60-inch chord as used in their bid, but removed the Fowler flap and tacked on a 24-inch chord external airfoil flap. In effect, this raised the actual wing area from 200 square feet to 280 square feet. Since the aerodynamic and structural analyses are calculated using coefficients (lift, drag, and moment) which are based on wing reference area and chord, the reference area and reference MAC were kept the same to keep from having to redo every analysis. Keeping with the same reference area and chord, when the actual area and/or chord change, is common in aircraft development. When tested in the wind tunnel, the original spoiler design was much more effective than the estimates, and as a result, the spoiler span was reduced in half.

The wind tunnel testing involved runs in determining lift, drag, stability, and control for both the landplane and floatplane configurations. Many design features

Model with the original single tail, but with the external airfoil flap in the neutral position (author's collection).

Model with the original single tail, but with the external airfoil flap in the full down position (author's collection).

such as wing dihedral and wing incidence were revised based on the wind tunnel tests.

An estimated drag buildup was made based on wind tunnel tests, other available Boeing data, and technical references. The drag estimate was based on the following assumptions:

1. The landing gear was unfaired.
2. The cooling drag was zero. The exhaust ejectors would supply sufficient energy to cool the engine.

Table 5.1. Estimated Zero Lift Drag (Landplane)[7]

Item	Parasite Area f, sq ft
Empennage	1.54
Main Landing Gear	3.00
Tail Wheel	0.50
Wing	2.90
Fuselage & Tail Boom	2.70
Cooling	0.00
TOTAL	**10.64**

Based on the wind tunnel tests, the seaplane (twin float) configuration estimated drag was 0.50 sq ft more than the landplane configuration.[8]

Interstate L-6 Testbed

The 102 horsepower Franklin O-200–5 engine was removed and a 125 horsepower Lycoming O-290–7 engine was installed on Boeing's Interstate L-6 testbed aircraft to aid in the power plant installation design of the XL-15. This enabled early flight tests and evaluation of some of the proposed XL-15 engine installation concepts and a check-out of the proposed propeller.[9]

During these L-6 flights, the two-position wooden Sensenich controllable propeller specified in the proposal was unsatisfactory at low speed, where it produced only about 70 percent of the estimated thrust. Other wooden Sensenich blades were also tested, and none were suitable from a performance standpoint. Boeing next tried a fixed pitch aluminum McCauley propeller with an activity factor of 80, which looked promising.

Boeing reported to the Air Force that metal blade propellers would be better than wood-bladed propellers. Next, Boeing tested a Beech Model R100 controllable pitch propeller with metal blades. They also looked at some McCauley aluminum blades which were made to fit the Beech hub.[10]

The L-6 served as a flying testbed in other areas besides propulsion. A new wing with 40 percent chord flaps and scoop-type spoiler ailerons was built and installed on the L-6 testbed to obtain preliminary flight test data for the XL-15 design.[11]

Mockup

The contract required a full-size mockup of metal and wood for inspection of design features. This mockup had operational flight controls and systems where possible. The mockup inspection at Boeing Wichita was from September 3 to 6, 1946, and attended by Wright Field civilian and military engineers, Army Air Forces officers, and Army Ground Forces officers.

The mockup had a single piece observer's door hinged at the top and rotated upwards against the bottom of the boom, but this configuration was rejected. The L-6 testbed was also shown at the mockup inspection and allowed detailed looks at the working engine installation.[12]

The Boeing engineers also used the mockup to evaluate different hinge arrangements for the pilot's door. They tried hinges on the front, rear, and top of the door. Door hinges mounted on the rear were chosen.

Table 5.2. Geometry Changes from Bid to Prototype

Model		451–2 Bid	XL-15 As Built
WING			
Span	inches	480	480
Mean Aerodynamic Chord (actual)	inches	60	84
Mean Aerodynamic Chord (reference)	inches	60	60
Area (actual)	sq ft	200	277
Area (reference)	sq ft	200	200
Airfoil		NACA 4415	NACA 4415
Leading Edge of MAC	FS	81.78	77.283
Incidence	°	3.5	2.5
Dihedral	°	4.0	1.0
Aspect Ratio (actual)		8.0	5.714

5. XL-15 Testing and Design Refinement

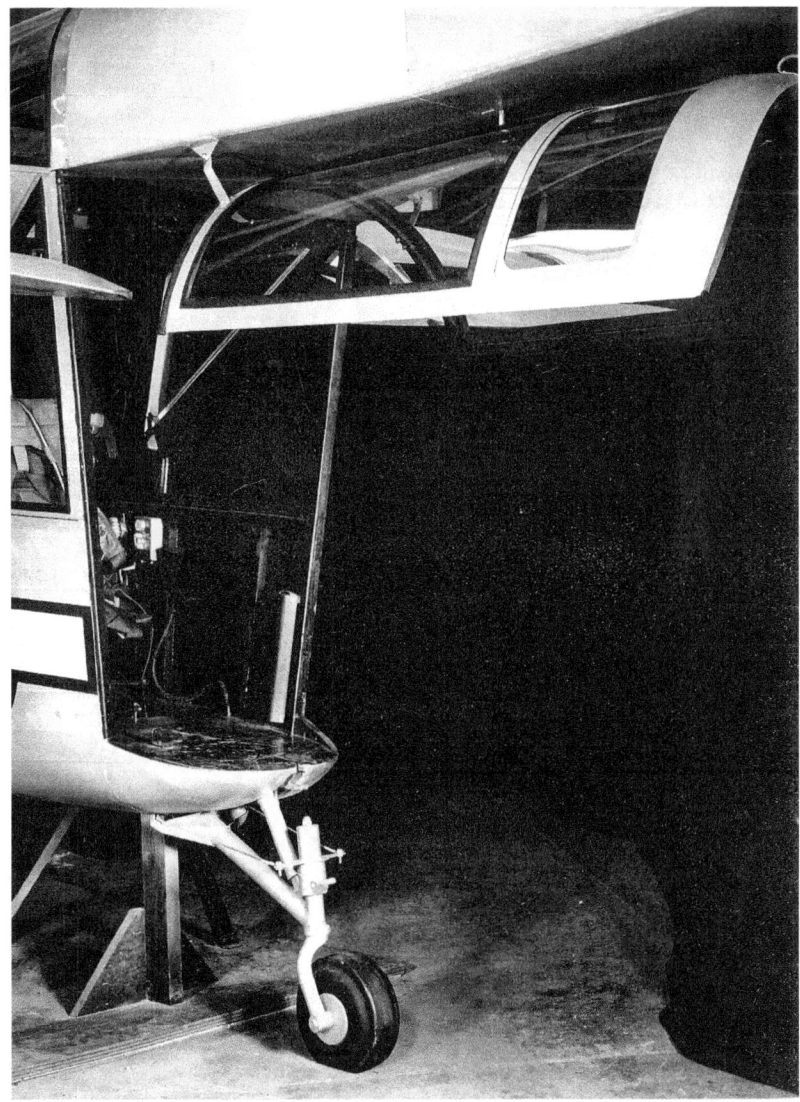

Mockup with a one-piece observer's door hinged at the top (author's collection).

Model		451–2	XL-15
Taper Ratio		1.0	1.0
FLAP			
Type		Fowler	external
Span (each)	inches	202	215
Chord	inches	24	24
Area	sq ft	67.33	71.67
Airfoil			NACA 23012
Inboard End	WS	24.3	16.4
Outboard End	WS	225.6	231.4
Deflection (min)	°	0	-10
Deflection (max)	°	40	35
AILERON			

Model		451–2	XL-15
Type		plain with horn	differential motion of flap
Span (each)	inches	12	215
Chord	inches	26.2	24
Area aft of hinge line (each)	sq ft	1.7	71.67
Horn Area ahead of hinge line (each)	sq ft	0.2	
Inboard End	WS	228	16.4
Outboard End	WS	240	231.4

Mockup with pilot's door hinged at the front (author's collection).

5. XL-15 Testing and Design Refinement

Mockup with pilot's door hinged at the rear (author's collection).

Model		451–2	XL-15
Deflection	°	± 25	± 4
SPOILER			
Type		circular arc	circular arc
Span (each)	inches	87	47.7
Projected Scoop Area (each)	sq ft	5.2	1.57
Chordwise Location	% c_w	66	
Hingepoint Location	% c_w	56.4	81.56
Inboard End	% $b_w/2$	60	71.7
Outboard End	% $b_w/2$	95	91.6
Deflection	°	60	65
Deflection (height)	inches	4.62	4.75

Table 5.3. Geometry Changes from Bid to Prototype (Continued)

Model		451–2	XL-15
		Bid	As Built
HORIZONTAL TAIL			
Span	inches	134	146.22
Root Chord	inches	50	39
Tip Chord	inches	35	39
Mean Aerodyamic Chord	inches	43	39
Area	sq ft	40	39.35

Mockup with pilot's door hinged at the top (author's collection).

Model		451–2	XL-15
Airfoil		inverted Clark Y	inverted NACA 4412 modified
Incidence	°	-1	-1.5
Dihedral	°	0	
Aspect Ratio		3.12	3.75
Taper Ratio		0.7	1
Tail Arm at Design C.G.	feet	16.7	16.67
ELEVATOR			
Type		plain with horn	nose balance
Span	inches	134	145.6
Root Chord	inches	20	17.7
Tip Chord	inches	18	17.7
Area aft of hinge line	sq ft	17.3	17.86

Model		451–2	XL-15
Horn Area ahead of hinge line	sq ft	1.1	2.9
Deflection (up/down)	°		30/20
Tab Type		plain	servo tab
Tab Span	inches	40	90
Tab Chord	inches	3.5	5.3
Tab Area	sq ft	0.97	3.2
Tab Deflection (up/down)			20/20
VERTICAL TAIL (each)			
Type		inverted	end plate
Number		1	2
Span	inches	41.35	54.96
Root Chord	inches	61	46.48
Tip Chord	inches	40	16.32
Mean Aerodyamic Chord	inches	49	33.6
Area	sq ft	14.5	11.15
Airfoil		BAC-66	NACA 0012 modified
Aspect Ratio		1	1.88
Taper Ratio		0.66	0.35
Tail Arm at Design C.G.	feet	16.7	16.61
RUDDER			
Type		plain with horn	plain
Span	inches	49.5	
Root Chord	inches	26	
Tip Chord	inches	21	
Area aft of hinge line	sq ft	7.2	3.17
Horn Area ahead of hinge line	sq ft	0.5	0.314
Deflection	°		± 25
Tab Type		fixed	fixed
FUSELAGE			
Aircraft Length	inches	322	314.89
Maximum Height	inches		61.9
Maximum Width	inches		48

Phase I Flight Testing

The Phase I flight test was the flight testing that was performed by the contractor (Boeing Wichita) on the first XL-15 airplane before it was turned over to the Army Air Forces for official government testing. The purpose of this testing was to verify the XL-15 was safe to fly, and the performance was satisfactory.

A safety and engineering inspection of the prototype was held at the Boeing Wichita plant on July 9, 1947, to clear the prototype for the flight test.[13]

The first taxi tests of XL-15 46–520 were made on July 12, 1947, at the Wichita Municipal Airport (now McConnell Air Force Base), where Boeing Wichita was located. During the initial ground power run-up, even with the elevators at full up, the tail wheel shock was fully extended, thus indicating an upload on the tail. This was countered by moving the flaps to the 10° full-up position. The taxi tests consisted of several runs downwind and upwind at speeds varying from 10 to 35 mph indicated. All flight controls appeared responsive and functioned satisfactorily.

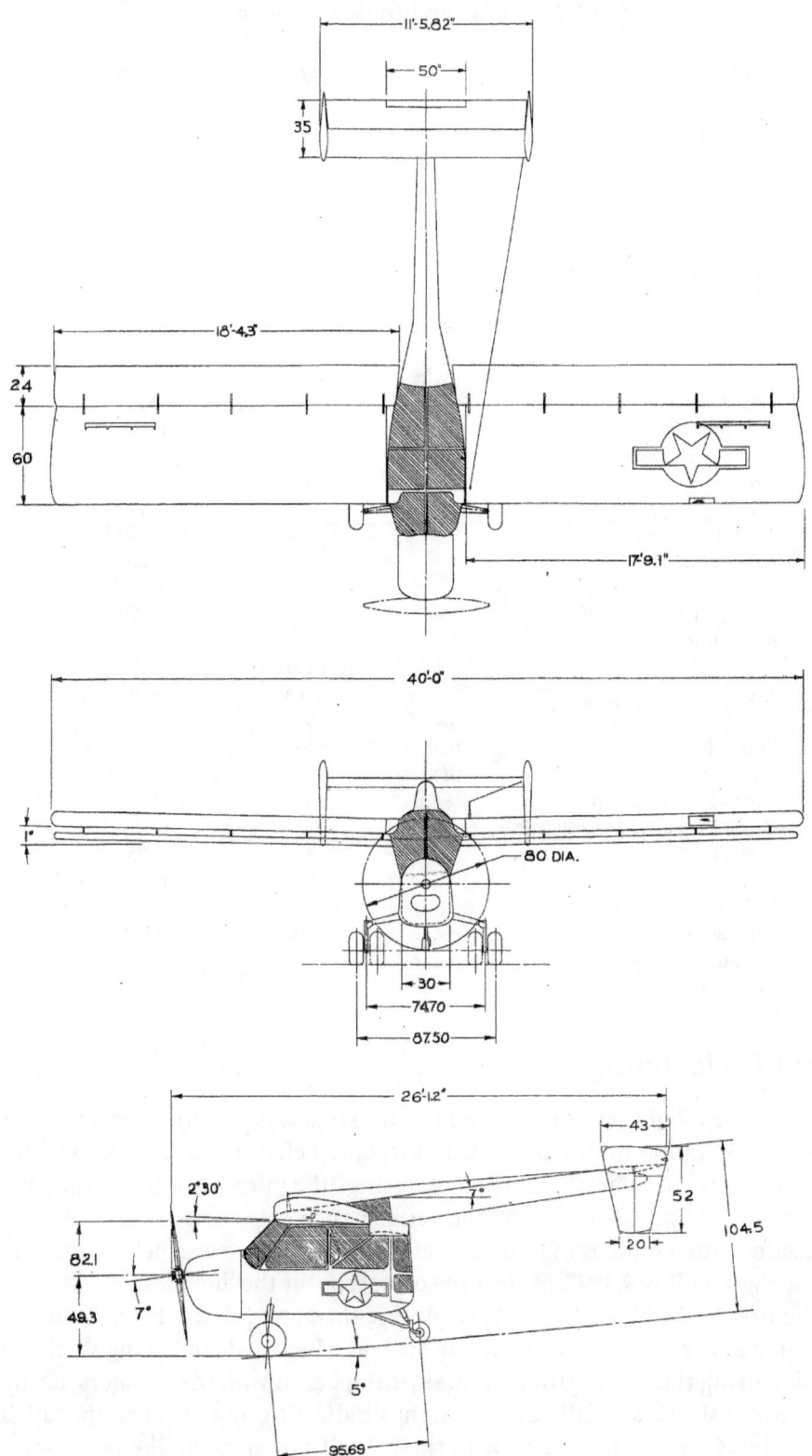

Three-view of the XL-15 as initially built (author's collection).

Brakes were somewhat soft and would not hold during the full-throttle run-up. The tail wheel steering was extremely stiff and thus made the airplane difficult to turn. Engine cooling during these ground tests appeared satisfactory as maximum head temperatures observed were on the rear No. 3 and 4 cylinders of 220° C. Except for the brakes and the stiffness in the tail wheel steering mechanism, the airplane was considered satisfactory for flight.[14]

The first flight of the XL-15 was made by Elton Rowley[15] the next day, July 13, 1947, just a little over a year after Boeing got the contract. Gross weight on the first flight was 1,800 pounds, and the CG was at 37.5 percent. A 10-minute crosswind ground cooling check was made at 1560 rpm with the propeller set at low pitch. The maximum cylinder head temperatures observed were No. 1–200° C; No 2–205° C; No. 3–215° C; No. 4–215° C. Maximum oil temperature observed was 85° C. The outside air temperature was 34° C (93° F).

The first takeoff was at 13:42 on runway 18 Left. The wind was about 20° from the right at 20 mph. During the initial takeoff run, power was held down so that the airplane would attain a speed just below takeoff with the flaps in the 10° up position. All controls appeared to be responsive, and the flaps were then deflected to the neutral position. The throttle was advanced to full open. The ship became airborne immediately. On the climb-out, the head temperatures began to climb rapidly. No. 3 cylinder exceeded the 260° limit about one and a half minutes after takeoff. At this point, power was pulled down as low as possible, No. 3 head appeared to stabilize at 265° C at low cruise power. On the base leg, the power was reduced, and No. 3 head dropped

Initial configuration of the XL-15 (AAHS—E. Stoltz).

slightly below 260°. The maximum oil temperature observed was 105° C. The flight was terminated due to excessive head temperatures. Landing was made at 13:51 on runway 18 Left with approximately 45° crosswind of 20 mph. No difficulty was experienced in handling the ship under these conditions.

An outstanding characteristic noted in turning flight was that very little rudder was required to execute a coordinated turn. All control forces, including flaps, appeared satisfactory. Also, no tail buffeting was noted. The pilot did note the tail wheel steering as being unsatisfactory. Rowley also recommended that ventilators should be added to the windshield and around the pilot's feet. There were also notes of various rattles, vibrations, and other minor discrepancies.[16]

A series of brief check flights were made to look qualitatively at the airplane's static and dynamic stability about all axis and that the aircraft had reasonable control forces. Stick-free decreases of 15 to 20 mph indicated airspeed resulted in the ship returning to normal airspeed in about two and one-half oscillations, indicating satisfactory longitudinal stability. There appeared to be no short-period oscillations. Lateral stability was considered normal for a light aircraft, with a tendency toward spiral instability. Well-coordinated turns could be made using ailerons alone. The elevator was extremely powerful, and when landing at maximum forward CG, there was no difficulty getting the tail down.[17]

Just over two weeks after the first flight on July 28, 1947, Elton Rowley flew the XL-15 to Fort Riley to demonstrate the airplane to the U.S. Army Chief of Staff, General Dwight D. Eisenhower. The demo consisted of a takeoff with a less than 100-foot ground roll, then climbing to 500 feet in less than 1,000 feet, followed by a 180° turn and a descent and landing with less than 200-foot ground roll. Another takeoff followed by a turn downwind then performed the rare attitude change without an altitude change that this aircraft was uniquely capable of, 25° nose up then 25° nose down while maintaining level flight. Next came a 150-foot radius 6g square turn. The demo ended with another maximum effort landing. Ike,

Elton Rowley demonstrating a short takeoff (author's collection).

who was himself a private pilot and had many flights in liaison airplanes, walked out to the XL-15 and told Rowley, "Good show. That kind of flying takes practice. This airplane is no Cub." Later Ike told Harold Zipp, Boeing Wichita Chief Engineer, "This airplane would take a lot of pilot training to get the full potential." The Boeing people thought of the L-15 as an easy and simple airplane to fly and were not expecting that kind of reaction from the general.[18]

There were approximately thirty follow-on flights aimed at fixing the engine cooling problem. Different baffle arrangements were tested with no measurable effect. There was a 40°C difference in head temperature between the front and rear cylinders. The cooling problem was finally cured by installing a "pepper shaker" nozzle in the carburetor, which improved the distribution to such an extent that the head temperature differential was reduced to 10°C, and the cooling was found to be satisfactory.

During this testing time, the GFE brakes kept failing, and Boeing had to install a set of Goodyear brakes to keep the flight test program going.[19]

The initial stall flight test on the XL-15 showed unsatisfactory stalling characteristics when the flaps were more than 10° down. These tests were run power off and had a roll to the left, which was pronounced and uncontrollable due to lack of rudder power. The neutral flap power-off stalls were satisfactory; however, there was a

The XL-15 could be flown at extremely high climb gradients (author's collection).

pronounced lack of stall warning. The left rolling tendency made it necessary to use full rudder by the time the stall occurred, leaving no rudder available to control the resulting roll. Aileron control against the roll appeared to advance the stall of the left wing. After the initial stall-check flight, spoilers (stall strips) were installed on the leading edge of the inboard section of each wing. The rudder area was increased 50 percent, which resulted in the vertical tail area being increased 15 percent, and the flap differential for aileron control was increased from 4 to 5°. This became the standard for the later YL-15.

The final stall test results were quite satisfactory. The aircraft appeared to be slightly unsymmetrical to the left for all flight conditions, but the roll was not violent, and the stall warning was good. The power-on stalls were considerably improved, with the airplane being slightly unsymmetrical to the left, but the roll was not violent, and there was considerable stall warning. It was possible to hold the airplane on a given heading using the rudder throughout the complete stall range. Aileron control was effective immediately before and after the stall but not at the initial pitch. The full flap indicated stalling airspeeds were 32 mph power off and 22 mph power on.[20] Spin entries were attempted at 10,000 feet pressure altitude from both power on and power off stalls and found very difficult to perform. The first was a left spin attempt with controls fully deflected with the spin; the airplane recovered after one and one-half turns. A right spin was attempted with the same result. This was followed by using several different procedures of entry and trying to hold the ship in the spin. The only one that had any real effect was the use of a quick burst of throttle to give power at the point where controls were applied to start the rotation. Ailerons with or against the spin appeared to make little difference. From the testing, it appeared that the XL-15 was incapable of prolonged spinning at normal CG location at any flap deflection. It also appeared the aircraft

The initial XL-15 right exhaust (author's collection).

would not spin unless the rudder was applied at the point of stall, not before or after. During the first turn of a spin with flap neutral, the indicated airspeed was slightly above the stall speed and increased rapidly after the turn, where self–recovery starts to occur. There were no vibrations or abnormal control pressures observed in any of these spin tests. The airplane attitude was nearly straight nose down at all flap settings, and with no back pressure on the stick.[21]

After the engine overheating was corrected, initial flight tests were disappointing from a performance standpoint. The 80-inch diameter Sensenich constant-speed propeller with wooden blades the airplane was originally equipped with having proved to be unsatisfactory for the tip speeds encountered at 2600 rpm and resulting in a rate of climb of only 405 fpm (guarantee was 690 fpm), Boeing tested ten propellers of various makes and types (fixed-pitch, two-position, and constant speed with wood, metal, aluminum, and plastic blades) before an 80-inch diameter McCauley fixed-pitch metal propeller having an activity factor of 80 was chosen as the definitive propeller.

Most of the propeller tests were flight tests, but low-speed thrust tests were also conducted by allowing the XL-15 airplane to pull an automobile at speeds (0 to 50 mph) controlled by the automobile driver using his brakes. The excess thrust was

The initial XL-15 left exhaust (author's collection).

recorded from a spring dynamometer in the tow line at five mph intervals. The aircraft frictional resistance and drag readings at these low speeds were determined by the reverse of this procedure with the automobile towing the airplane. These tests were done early in the morning when there was zero wind.[22]

On the morning of August 12, 1947, with the McCauley Model 1A170-LE042 fixed-pitch metal propeller installed, Boeing began a series of field performance tests. Two takeoffs and landings were made, and in the process of making the second landing, the XL-15 nosed over, damaging the propeller, engine, and engine mount. This flight was also the first time in which 40 degrees of flap had been used for landing. The accident was stated to be the result of several factors, including the first use of 40° flap setting, the need to develop the technique of landing with full down flaps, the grabbing of the temporary Goodyear brakes, and the stiffness of the tail wheel shock absorber strut. The aircraft was repaired and ready for flight within two days.[23]

Even with the definitive propeller, performance was still only a 475 fpm climb at sea level and an 11,000-foot service ceiling. It was felt that there was no more to gain from the propeller, so the power plant installation came under deep review. The aircraft was initially equipped with exhaust pumps in anticipation of low-speed cooling problems. There were some cooling problems found, but these turned out to be due to improper fuel mixture distribution and were corrected by a change in the carburetor main nozzle. The exhaust pumps did contribute to cooling, but their

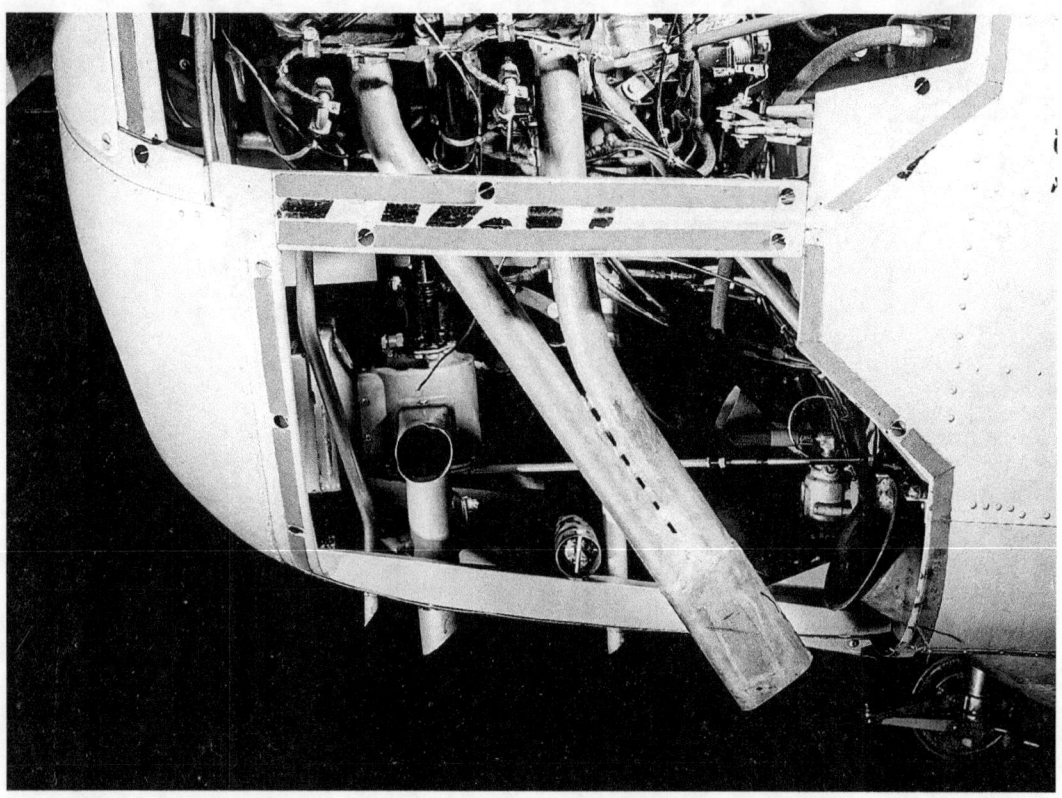

An exhaust pipe configuration which was tested (author's collection).

contribution was no more than could be obtained by using a simple cowl flap on the engine cowl lower surface. The exhaust pumps were removed and were replaced by conventional exhaust manifolds, each collecting the exhaust gas from two cylinders based on the engine manufacturer's recommendation. The result was no change in the rate of climb.

At this point, it was believed the low rate of climb was most likely due to power loss. A series of ground run tests were conducted using a test club propeller on various manifold configurations, and a manifold arrangement with tuned individual stacks for each cylinder was the chosen configuration. The rate of climb for the tuned pipes was 640 fpm with a service ceiling of 12,750 feet.

During these climb tests, it was extremely difficult to correctly set the mixture for best power at altitudes above 8,000 feet. The mixture control quadrant was marked off in 16 segments of one-quarter inch each. Flight test showed that a linear mixture control adjustment of six segments (0 to 1.5 inches) was needed when climbing from 1,000 to 8,000 feet. But from 8,000 to 12,000 feet the required adjustment was only one-quarter inch segment. The control motion between the cockpit mixture control lever and that on the carburetor was linear. It was pretty apparent that the entire carburetor lever movement was not required for altitude mixture adjustment; still, the whole movement was needed so that the idle cut-off could be used.

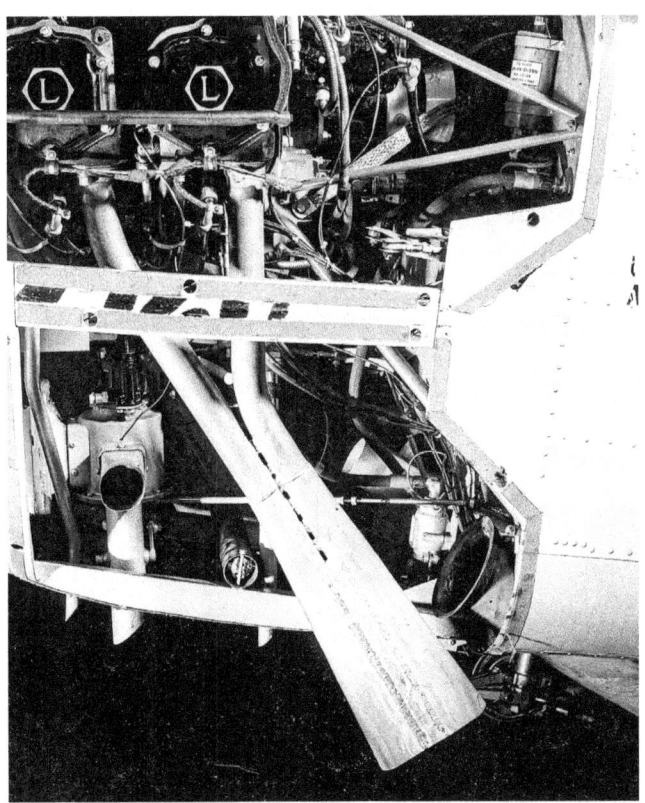

Another exhaust pipe configuration that was tested (author's collection).

Since it was extremely difficult to lean for best power above 8,000 feet using marks on the mixture control, flight tests were made in which the engine was leaned to pre-determined fuel flows. This required a fuel flow meter to be installed in the test aircraft cockpit, and the engine was leaned to a predicted fuel flow instead of to a maximum rpm increase, resulting in a service ceiling increase of 1,400 feet.

Lycoming was requested to supply Boeing Wichita with an automatic fuel control for the carburetor. Lycoming agreed but found that none of the available automatic fuel control units could work with the special "pepper shaker" nozzle carburetor Boeing was using on the XL-15. Lycoming ultimately shipped an experimental altitude compensating carburetor to Boeing. This new carburetor produced a definite improvement in a climb up to 8000 feet, but above that, power fell off, and there was no climb improvement. Also, with this carburetor, the engine cooling problem returned.[24]

The dates for the propeller and engine power testing described in the paragraphs above are not known by the author, and while some of the testing occurred during the Phase I testing, much of it probably happened during the break in Phase II testing. (Phase II testing was the performance testing done by Army Air Forces test pilots and engineers on the XL-15 and is described in Chapter 6. Phase II testing started on September 3, 1947, and was prematurely shut down on October 27, 1947, because the XL-15 could not meet guarantees, and the AAF told Boeing to get their act together. Boeing Wichita immediately launched into a performance improvement program. Phase II testing was resumed by the AAF on January 15, 1948.)

As a requirement of Phase I testing, two preliminary structural integrity flights were flown. For these flights, a V.G. (Velocity-Gravity acceleration rate) recorder was installed on the forward face of the front spar which was located approximately on the CG of the aircraft (41 percent MAC). A visual accelerometer was installed on the fuselage structure just above the pilot's instrument panel, which was 36 inches ahead of the position of the V.G. recorder. The first structural integrity flight was made on August 20, 1947, to determine the accuracy of the visual accelerometer and to check the V.G. recorder for proper dampening. During this flight, two pull-ups were made at approximately 2 g's at 90 and 120 mph IAS. The aircraft returned to base, and the V.G. recorder was checked, and the visual accelerometer was noted to read approximately 0.2 G higher than the V.G. recorder. The requirement for the final flight was a 2.4 g pull-up; therefore, a 2.6 g was required on the visual accelerometer. The last preliminary structural integrity flight was made on the same day, shortly after the first flight. The airplane was loaded to 2,065 pounds at takeoff with the CG at 41 percent MAC. A climb was made to 5,000 feet and the throttle set to cruise rpm, and the aircraft nosed over into the dive. At approximately 125 mph IAS the pull-out was started. The maximum airspeed attained in the dive was 130 mph IAS. A reading of 2.7 g's was observed on the visual accelerometer at 115 mph IAS. After landing, the V.G. recorder was checked, and the trace was found to be clear and properly damped with a reading of 2.5 g's for the dive.[25]

Flight Test Instrumentation

The XL-15 had flight test instrumentation much more sophisticated than that customarily used on light aircraft and was probably more extensively tested than any previous airplane of its type and size. Flight test data were recorded using a special photo recorder incorporating an altimeter, tachometer, manifold pressure, two airspeed systems, and a clock with a sweep second hand. The photo recorder could be set to record at any desired time interval. An airspeed bomb on a trailing cable was used for airspeed calibration. For takeoff and landing performance tests, there was a light mounted in the aircraft's window, which was electrically connected to the photo recorder and flashed with each film frame exposed. The window light was also filmed on a field performance camera set up 1,500 feet on the side of the runway. Thus, it was possible to calibrate the field camera to the clock in the photo recorder and coordinate the onboard photo recorder data with the aircraft position data that was geometrically determined from the field camera.

A zero-thrust glide polar was determined in the XL-15 flight tests using a unique zero-thrust indicator. This indicator consisted of feeler actuating contact points against a ground metal plate mounted on the back of the propeller. The contact points closed a circuit that turned a light on in the cockpit. There was a 0.015-inch end play in the crankshaft of the Lycoming engine, and at the zero-thrust condition, there was a constant making and breaking of the contact as the propeller moved back and forth. When the light was flashing in the cockpit, the test pilot knew he was at zero thrust. The glide tests also required the development of a special angle of attack (AOA) indicator. Calibration of the angle of attack indicator was done in level flight using a pitch indicator. AOA was measured in both level flight and glide conditions and used to develop lift versus angle of attack curves. The zero thrust flight test polar matched the final wind tunnel results for the XL-15. Also, using the zero thrust glide polar and the speed power test data confirmed the propeller curves being used.[26]

Test instrumentation on pilot's panel (author's collection).

Test instrumentation in observer's position (author's collection).

Phase III Flight Testing

As a result of a conference with the Army and Boeing Wichita representatives, it was decided that the second airplane (46–521) would be retained at the Boeing Wichita plant for Phase III changes and flight testing instead of sending the aircraft to Eglin Field for tactical utility tests. An additional $9,565.34 was added to the XL-15 contract for the Phase III Tests.[27]

The maximum rates of roll were measured at various airspeeds and flap deflections for the standard aileron-spoiler configuration and again with the spoilers extended by a simple 8-inch span split-flap at each end of both the right and left-hand spoilers, providing a 30 percent increase in spoiler span. The tests were run to determine the possibility of improving roll control at low airspeed. The standard roll control was satisfactory based on wingtip helix angle (pb/2V), but since the stall speed was so low, the actual rolling velocity is lower than desirable at landing velocity. The test data indicated:

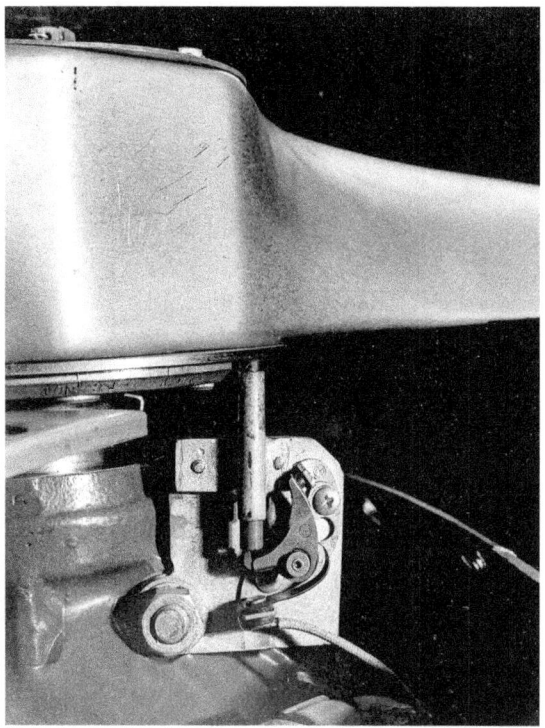

Zero-thrust indicator contact (author's collection).

(a) Roll control at normal speeds was good.

(b) The rolling control decreased rapidly as the stall was approached.

(c) Roll control at a given speed improved with increased flap angle.

(d) Roll control with flaps up was not appreciably affected by the spoiler extension.

(e) Roll control with 20° degree flaps was improved approximately 20 percent at lower speeds by the spoiler extension.

(f) Roll control with full flaps was improved approximately 25 percent at lower speeds by the spoiler extension. However, this rather large improvement was not noticed by the flight crew due to the low absolute rate of roll and the lack of symmetry.

(g) The aileron spoiler should be doubled in span for future airplanes of this type.[28]

Because of the full-span flaps, the aircraft's normal landing is 35 mph, and light-weight landings were recorded at as low as 25 mph in flight tests. The power effects on the stall are such that the power on $C_{L_{max}}$ is double that of power off. Thus, lateral control at low airspeeds is very important. The aircraft in landing configuration at 1.10 power-off stall speed has a pV of 10 feet per second and pb/2V of 0.083. The Air Force specification requirements were a pV of 10 feet per second and a pb/2V of 0.070. Boeing felt that the Air Force requirements were satisfactory for conventional airplanes but that they should be higher for low-speed aircraft like the XL-15. Because of the combined use of differential flaps and spoilers as ailerons, the yaw in roll is neutral. The maximum rate of roll with flaps full down is 25 degrees per second. There is no appreciable difference in roll characteristics between flaps zero or full down.[29]

The Phase I flight testing had revealed the stall characteristics to be very poor and the rudder control weak. It was decided to expand onto the limited Phase I stall strip testing by testing stall strips of several sizes and in multiple locations to improve the stall characteristics. XL-15, A.F. Serial No. 46–521, flew 15 flights testing different stall strip configurations, along with five flights made on YL-15, Serial No. 47–423. In addition, on later flights, the rudder travel was increased from 25 to 35 degrees. which provided more rudder feel, and the rudder control was improved at the full flap stall.[30]

Airspeed probe and angle of attack vane (author's collection).

After this considerable experimentation of fixes, the full power full flap stall was as gentle as the power off flap zero stall. This resulted from using tailored stall strips located approximately one-fifth of the semi-span outboard of the body and having a length of 10 percent semi-span. These stall strips improved stall control and resulted in a lower stall speed instead of the expected higher stall speed. Elevator power was also an effect in achieving low stall speeds and it was found necessary to use a slotted elevator. Army Air Forces tests showed the full power full flap stall speed was 28 to 30 mph.[31]

The first XL-15 airplane (46–520) was used on a test flight to determine the frequency, amplitude, and other characteristics of flap vibrations that had been reported to occur during a full-power climb. The frequency was measured using a General Radio Strobotac. Amplitude was measured by comparing it to a scale that could be viewed simultaneously with the trailing edge of the flap. The flap only appeared to vibrate at one condition, full throttle climb. The right flap was observed to vibrate at 2,440 cycles per minute (cpm) with an amplitude of 1/16 inch, and the right flap vibrated at 2,430 cpm and 1/32-inch amplitude while the engine tachometer read 2410 rpm. On another occasion, the right flap vibrated at 2,400 cpm with an amplitude of 3/32 inch, and the right flap vibrated at 2,400

Slots were tried on the outboard wings (author's collection).

cpm and 1/16-inch amplitude while the engine tachometer read 2380 rpm. No vibration could be detected at any other rpm. Speeds up to 120 mph were investigated. This particular flap vibration condition had never been encountered on any other L-15 aircraft, and no flap vibrations or flutter were indicated during the dive tests.[32]

The Air Materiel Command requested Boeing to conduct tests on a modified manual control Marvel-Schebler carburetor. The modification was a rework to provide a more linear mixture control adjustment at altitude with a reduced sensitivity such that *any* pilot should be able to operate it. Before it was installed on XL-15 (46–521), two sawtooth climbs were made to compare to the Phase I data from XL-15 (46–520). The two airplanes were found to have the same climb speed performance. The modified carburetor was installed on 46–521 and two continuous climbs made to the aircraft's service ceiling. From the standpoint of adjustment of the mixture above 8,000 feet, the modified carburetor was a definite improvement over the original unit; however, the engine's power output seems to have been reduced slightly with the modified carburetor. This change could have been due to the change in jet size from 975 cc to 1,000 cc. Also, it was noted that the automatic fuel control carburetor tested during Phase I was superior to both the modified and the original in power output. The Phase III summary report stated, "It is apparent that further improvement of the carburetor is possible."[33]

The structural integrity flight demonstration was done per Wright Field requirements. This is the infamous "see if you can pull the airplane apart" series of tests. There were eight conditions tested:

    a. Push-down.

Requirement: $n = -2.0$ g at CAS = 112 mph. Pressure altitude not to exceed 10,000 feet.

    Result: $n = -2.3$ g at CAS = 124 mph at altitude of 6,460 feet.

    b. Pull-up.

Requirement: $n = 4.0$ g at CAS = 88 mph CAS. Pressure altitude at pull-out optional.

    Result: $n = 4.0$ g at CAS = 88 mph.

    c. Pull-up.

Requirement: $n = 4.0$ g at CAS = 112 mph. Pressure altitude not to exceed 10,000 feet.

    Result: $n = 4.1$ g at CAS = 105 mph at altitude of 5,590 feet.

    d. Dive .to obtain CAS = 200 mph. Pull out.

Requirement: $n = 2.47$ g at CAS not less than 10 mph below maximum CAS obtained in dive. Pressure altitude for obtaining required airspeed and acceleration not exceeding 10,000 feet.

    Result: $n = 2.5$ g at CAS = 209.5 at altitude of 3,250 feet.

    e. Pull-up with flaps down 25°.

Requirement: $n = 4.0$ g at CAS = 80 mph.

Result: $n = 3.9$ g at CAS = 81 mph at 4,840 feet.

    f. Abrupt deflection of flaps to 25°.

Requirement: CAS = 80 mph. Calculated value of n is 2.87 g

Result: $n = +2$ g at CAS = 80 at altitude of 4,990 feet.

    g. Pull-up with flaps up at –10°.

Requirement: $n = 4.0$ g at CAS = 112 mph.

Result: $n = 4.3$ g at CAS = 111.5 mph at 4,800 feet.

    h. Abrupt roll from a steady left turn.

Requirement: $n = 2.67$ g (angle of bank approximately 68°). Airplane must roll to at least same degree bank in opposite direction.

Result: $n = 2.8$ g at CAS = 77 mph at altitude of 7,030 feet. Airplane roll to the right, 90°; airplane roll to the left, 75°.

The test pilot pointed out that there could be a dangerous condition for inexperienced pilots at airspeeds above 125 mph, such as when on aerial tow. This is because at these high speeds, the control forces are extremely low, and it was recommended that a bellows device be installed in the control system to increase stick pressure at high speeds to prevent rapid stick movement.[34]

During the Phase III testing, additional spin tests were made using the second XL-15 aircraft. This was done since after the original Phase I spin tests the XL-15s had been modified with a reverse slotted elevator, more rudder area with 10° more rudder travel, and more effective outboard spoiler (stall) strips.

With 0° flaps, the spin entry characteristics were considered excellent. From a stalled condition, there was very little tendency to enter a spin inadvertently. However, a spin could be easily entered by applying full rudder in the desired spin direction at any time, either just before the stall, during, or after the stall break. Entries were attempted from various combinations of power and turns, and it was found that a clean spin could be entered by positive rudder use from any condition in which a stall could be attained. The characteristics in the spin were considered to be good. The rotation was on the high side, but there were no abnormal or undesirable characteristics, and this was true even when lateral control was applied in the spin. The airplane had to be held in the spin by use of full rudder in the direction of the spin and full aft stick; any relaxation of either control and the airplane would either recover rapidly or go into a spiral. If held into a prolonged spin, the aircraft became stable in two to three turns, with the rudder forces being moderate and forward stick forces becoming high. Altitude lost from entry to recovery was not excessive for this class of aircraft. Applying full reverse rudder followed by stick forward to neutral resulted in smooth recoveries from six turn spins in one-eighth to one-quarter turn (the requirement was one and a half turns). Many other recovery control movements were tried, such as bringing controls to neutral position only, reversing elevator before rudder, and using only elevator or only rudder to uncover any dangerous or undesirable recovery characteristics; the airplane fully recovered within a maximum of a five-eighths turn, regardless of the control technique used or the number of turns preceding the recovery. It was believed that it would be extremely difficult, if not impossible, to use the controls in such a way that a rapid recovery would not result.

With 20° and 35° flaps, attempted spins indicated the aircraft had strong spin-resistant characteristics at those flap settings. The spin entry would be normal, usually accompanied by a rapid initial nose drop, but at about the ½ turn point, the spin degenerated into a steep spiral from which immediate recovery could be made. Experience indicated that it would be very difficult to enter a spin with flaps at 20° or greater from a steep turn. This condition is considered highly desirable in that spins from the normal landing configuration would not be as dangerous as in many other light aircraft.[35]

Other Tests

Tests were also made on the XL-15 with both floats and skis. There is a photo of an XL-15 with Brodie gear, but the author has found no records of it being flight tested.

Most of this flight testing was done at Boeing Wichita, but the aircraft were also ferried to Wright Field for tests by the Air Materiel Command.

In April 1948, Charles Lindbergh visited Boeing Wichita to get a demonstration of air to air refueling. While he was at the plant, he also flew in the XL-15. With

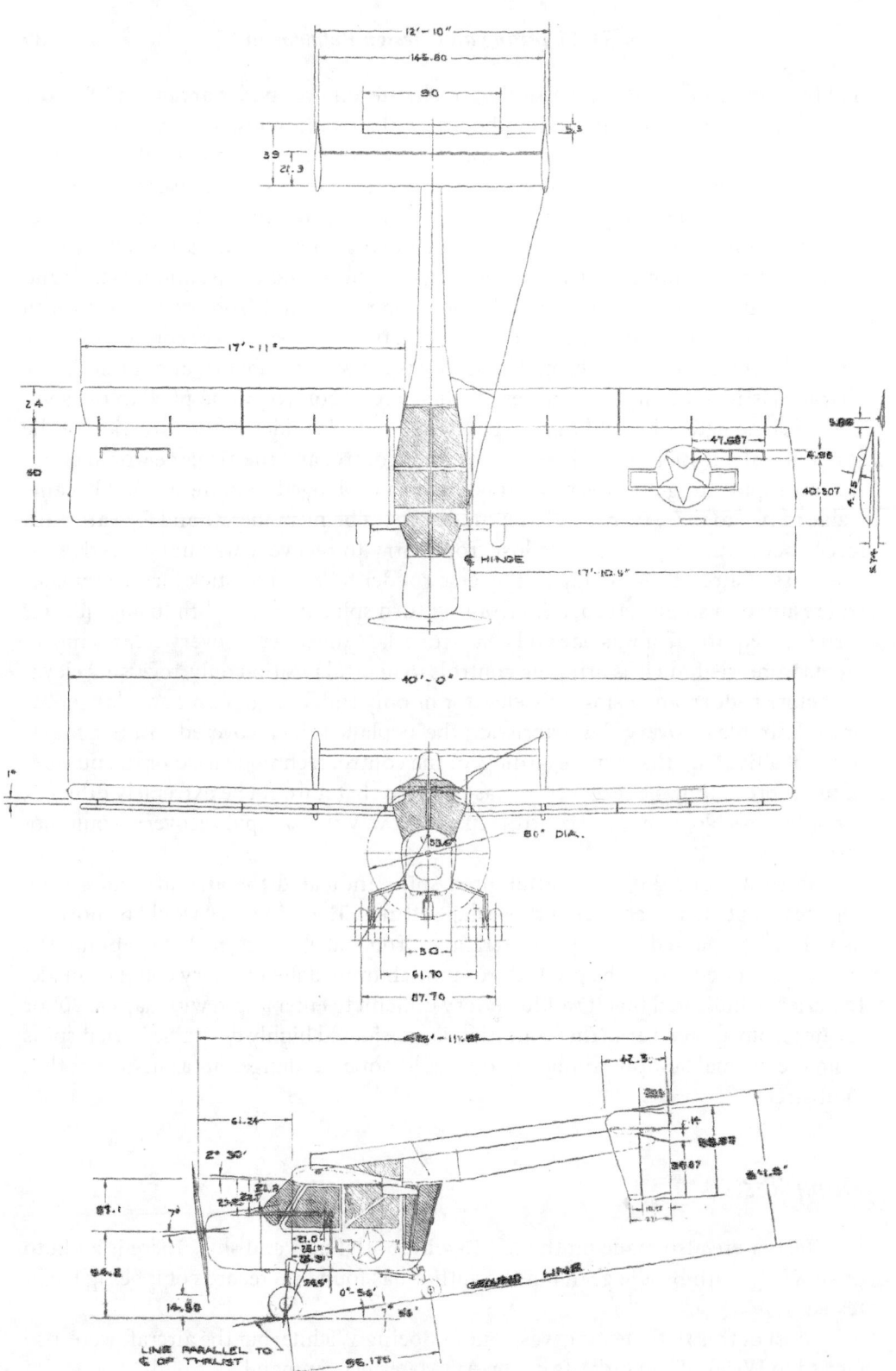

Three-view of the XL-15 in its final configuration (author's collection).

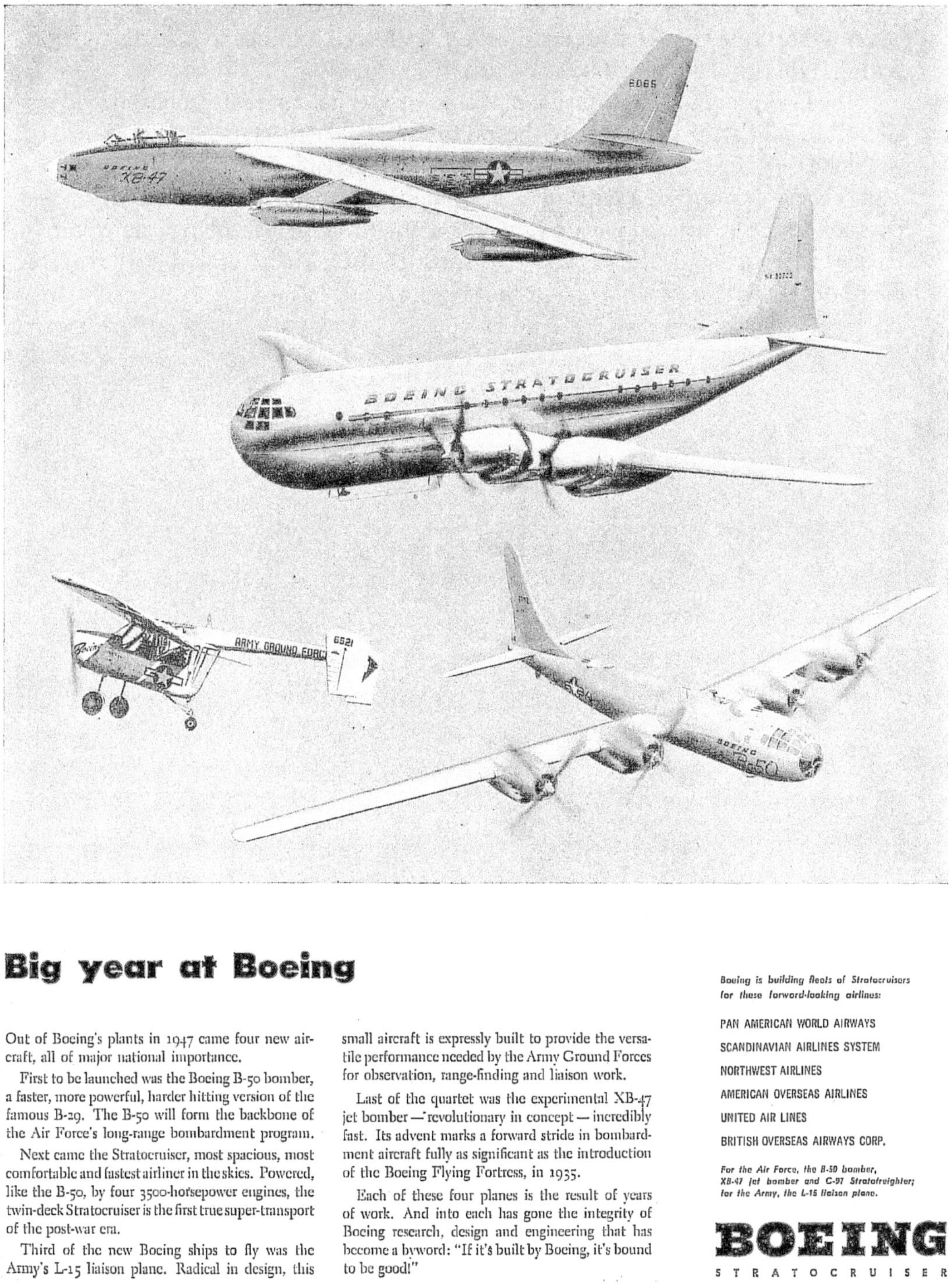

5. XL-15 Testing and Design Refinement

Big year at Boeing

Out of Boeing's plants in 1947 came four new aircraft, all of major national importance.

First to be launched was the Boeing B-50 bomber, a faster, more powerful, harder hitting version of the famous B-29. The B-50 will form the backbone of the Air Force's long-range bombardment program.

Next came the Stratocruiser, most spacious, most comfortable and fastest airliner in the skies. Powered, like the B-50, by four 3500-horsepower engines, the twin-deck Stratocruiser is the first true super-transport of the post-war era.

Third of the new Boeing ships to fly was the Army's L-15 liaison plane. Radical in design, this small aircraft is expressly built to provide the versatile performance needed by the Army Ground Forces for observation, range-finding and liaison work.

Last of the quartet was the experimental XB-47 jet bomber — "revolutionary in concept — incredibly fast. Its advent marks a forward stride in bombardment aircraft fully as significant as the introduction of the Boeing Flying Fortress, in 1935.

Each of these four planes is the result of years of work. And into each has gone the integrity of Boeing research, design and engineering that has become a byword: "If it's built by Boeing, it's bound to be good!"

Boeing is building fleets of Stratocruisers for these forward-looking airlines:

PAN AMERICAN WORLD AIRWAYS
SCANDINAVIAN AIRLINES SYSTEM
NORTHWEST AIRLINES
AMERICAN OVERSEAS AIRLINES
UNITED AIR LINES
BRITISH OVERSEAS AIRWAYS CORP.

For the Air Force, the B-50 bomber, XB-47 jet bomber and C-97 Stratofreighter; for the Army, the L-15 liaison plane.

BOEING
STRATOCRUISER

Boeing bragged in this magazine ad about flying four new airplanes in 1947 (author's collection).

Lindbergh in the observer's seat, Rowley flew the demo that he had flown the year before at Fort Riley for General Eisenhower. Lindbergh declined to take the controls, stating, "This kind of flying takes practice."[36]

The L-15 program was unusual in that there was no separate structural load test aircraft to validate the strength of the aircraft. A structural static test of the aircraft was done very late in the program at the Wright Field Aircraft Structural Laboratory using the first prototype XL-15 46–520, which was transferred from the Flight Test Division to the Structural Lab on July 12, 1948, with static tests being carried out in the period from August 16, 1948, to November 17, 1948. The aircraft met all requirements of the static tests with a few minor reinforcements.[37]

6

Performance

Performance is the standard measurement of how well an aircraft does its prescribed job. USAF reports give minimum distances, not average distances, for takeoffs and landings.

Request for Bid

The ATSC's February 4, 1946, Request for Bid (RFB) for a field artillery observation airplane listed both the minimum and desired performance items.[1]

Table 6.1. RFP Requirement

Requirement		Minimum	Desired
TO dist over 50 ft—sod	ft	600	300
Landing dist over 50 ft—sod	ft	600	300
Landing Speed	mph	35	25
Max R/C	fpm	600	800
Service Ceiling	ft	15,000	18,000
Cruise at 75% Power	mph	90	
Range, with aux tank	s. miles	500	
Endurance, standard fuel	hours	3.5	4

Model 451-2 and XL-15 Estimated Performance

The guaranteed performance in Boeing's Model 451-2 bid dated March 6, 1946, which won the XL-15 competition, more than met all of the minimum required performance that the RFB specified. This performance estimate was based on standard performance estimation techniques of the time and was documented in a 202-page report.[2]

The design refinement Boeing undertook after being informed they were the competition's winner and building a testing a wind tunnel model was described in Chapter 5. A new performance estimate was made based on the changed geometry, the wind tunnel tests, and flight testing of the modified L-6 testbed aircraft.[3] This round of estimated performance was slightly reduced, but still met the minimum requirements and was accepted by the Air Materiel Command as the contract's official guaranteed performance.

A month before the first flight in July 1947, Boeing made another performance estimate based on the prototype's configuration.[4] This estimate was prepared by

Top: Boeing XL-15 sn 46–521 (author's collection). *Bottom:* Boeing XL-15 sn 46–520 (author's collection).

Top: Boeing XL-15 sn 46–521 (author's collection). *Bottom:* Boeing XL-15 sn 46–520 (author's collection).

Boeing XL-15 sn 46–521 (author's collection).

aerodynamics engineer Charles M. Seibel.[5] This refined estimate showed a slight climb performance increase but predicted the guaranteed takeoff distance and landing speed would not be met.

The first *Pilot's Handbook and Flight Operating Instructions for the XL-15 Airplane* was prepared by Boeing and, based on alculations, the presented performance, except for landing distance and cruise speed, met or exceeded the guaranteed performance.[6]

Table 6.2. XL-15 Estimated Performance

Requirement		Minimum	Bid Guarantee	Contract Guarantee	Pre-1st Flt Estimate	Pilot's Handbook
Weight	lbs		2,050	2050	2,050	2,050
TO dist over 50 ft-sod	ft	600	595	600	670	595
Landing dist over 50 ft-sod	ft	600	517	517	530	530
Landing Speed	mph	35	35	35	36.5	36.5
Max R/C	fpm	600	690	628	692	692
Service Ceiling	ft	15,000	17,800	16,400	16,800	

Requirement		Minimum	Bid Guarantee	Contract Guarantee	Pre-1st Flt Estimate	Pilot's Handbook
Cruise at 75% Power	mph	90	109	99	92	98
Range, with aux tank	s. miles	500	560	540	650	700
Endurance, standard fuel	hours	3.5	4	4	4.7	
Reference		RFP	WD-12056	WD-12102	WD-12130	WD-12174
Date		4-Feb-46	6-Mar-46	1-Oct-46	12-Jun-47	1-Aug-47

XL-15 Flight Test Performance

Up to the time that Boeing Wichita Chief of Flight Test Elton H. Rowley made the first flight of the XL-15 on July 13, the aircraft's performance was all estimated. Now the actual performance would become known. However, as is usual in flight testing a new aircraft, the early flights by Boeing were to verify the airplane was safe to fly in the operating envelope. Thus, Boeing concentrated on making sure the systems worked and on stability and control testing. If Boeing did any performance flight testing in the six weeks after the first flight it was probably just a token look.

The first real look at the performance of the XL-15 was in Phase II flight tests performed by the USAF—the new service branch created in July—in Wichita between September 3 and October 27, 1947, to determine if the aircraft met its guaranteed performance. These tests in serial number 46–520 were flown by USAF Captain Wilbur E. Fleck of the Air Materiel Command Cargo, Training, and Miscellaneous Aircraft Section. The test data were recorded and reduced by Charles H. La Fountain, an engineer in the Performance Engineering Section of the AMC Test Engineering Section. The aircraft test weight with test instrumentation, crew, and full fuel was 2,068 pounds at 41.8 percent MAC.[7]

When the XL-15 Phase II flight program began, Elton Rowley realized at last what General Eisenhower was talking about that day at Fort Riley. Captain Fleck had difficulty flying the XL-15 and only made it through after hours of briefing and instruction in flight. Rowley finally admitted to himself that the airplane was difficult to fly, particularly when a pilot was trying to meet the required Army level of performance.[8]

The majority of the flight tests were flown with a forged aluminum fixed pitch McCauley model 1A170–18044 propeller. Tests were also flown with a wooden blade constant-speed Sensenich Brothers model D-380H3 propeller with a high pitch limit of 15°7' and a low pitch limit of 6°5'. With the Sensenich propeller, the aircraft was three knots slower than with the McCauley propeller. Rate of climb was also much lower with the Sensenich than with the McCauley propeller.

Power on and power off stalls performed in the cruise, climb, and glide attitudes had virtually the same stall warning and stall break. The approach to a stall was evidenced by a slight airframe vibration and a noticeable lateral instability. The actual stall was a mild, straight-ahead, fall-through type. Power off indicated stall speeds were flaps up 47 mph, flaps 20° 38 mph, and flaps 35° 34.5 mph.

The Air Force flight test was discontinued before completion because the aircraft's performance was lower than that estimated and guaranteed by Boeing. The Air Force had flown the aircraft for approximately 30 hours in this initial testing.[9]

Table 6.3. USAF 1st XL-15 Flight Test Performance

Requirement		Contract Guarantee	1st USAF Flight Tests
Weight	lbs	2,050	2,068
TO dist over 50 ft—sod	ft	600	605
Landing dist over 50 ft—sod	ft	517	not tested
Landing Speed	mph	35	not tested
Max R/C	fpm	628	470
Service Ceiling	ft	16,400	11, 000
Cruise at 75% Power	mph	99	84
Range, with aux tank	s. miles	540	not tested
Endurance, standard fuel	hours	4	3.2
Reference		WD-12102	MCRFTP-2120
Date		1-Oct-46	5-Jan-48

This discontinuance of the testing by the Air Force was to allow Boeing some time to fix the performance problem. Boeing, due to its previous propeller testing, felt they had the right propeller. They believed the problem was due to a loss in engine power and started working on that avenue. The Boeing power loss work was described in Chapter 5.

The Air Force resumed their Phase II performance testing in Wichita on January 15, 1948, and flew 14.5 hours, ending the testing on January 23, 1948. Charles H. La Fountain was again the engineer, but the pilot for this round of testing was USAF Lt Colonel Robert R. Shaefer of the Air Materiel Command Cargo, Training, and Miscellaneous Aircraft Section.

For this second round of Air Force tests, the aircraft test weight with test instrumentation, crew, and full fuel was 2145 pounds with the center of gravity at 43.3 percent MAC. The propeller used was a forged aluminum fixed pitch McCauley model 1A170–18044 propeller with a blade setting of 13.25° at 0.75 radii.

The performance was definitely improved, but it still did not meet guaranteed cruise speed, max rate of climb, service ceiling, endurance, or ferry range. In the report table (see Table 6.4), the range was given as 414 statute miles at 102 mph (i.e., max power), while in another section of the report, it stated the maximum ferry range with an external tank was 576 statute miles at 81 mph, which would have met the guarantee.[10]

Table 6.4. USAF 2nd XL-15 Flight Test Performance

Requirement		Contract Guarantee	2nd USAF Flight Tests
Weight	lbs	2050	2090
TO dist over 50 ft-sod	ft	600	600
Landing dist over 50 ft-sod	ft	517	330
Landing Speed	mph	35	34.5

6. Performance

Requirement		Contract Guarantee	2nd USAF Flight Tests
Max R/C	fpm	628	600
Service Ceiling	ft	16,400	12,400
Cruise at 75% Power	mph	99	94
Range, with aux tank	s. miles	540	414 or 576
Endurance, standard fuel	hours	4	3.25
Reference		WD-12102	MCRFTP-2120 (Addendum 1)
Date		1-Oct-46	6-Feb-48

In the spring of 1948, XL-15 s/n 46–520 was at Wright-Patterson Air Force Base (as the former Wright Field would now be known), where it underwent Phase IV flight tests by the USAF to check engine cooling, additional comparative propeller tests, and longitudinal stability and control evaluation for a total flight time of 21 hours 36 minutes. The tests were piloted by USAF 1st Lt Robert L. Northrup of the Air Materiel Command Cargo, Training, and Miscellaneous Aircraft Section. The test data were recorded and reduced by Thomas J. Borgstrom, an engineer in the Performance Engineering Section of the AMC Test Engineering Section. Performance flight tests were measured with both a forged aluminum fixed pitch McCauley model 1A170–18038 propeller with a blade setting of 13.25° at 0.75 radii and with a Hartzell model 8031–64044–1L automatic two-position propeller. The Hartzell propeller was designed to change pitch automatically at a speed slightly above the best climbing speed of the airplane. For these tests the automatic feature was inoperative and the takeoffs and climbs were flown with the propeller in low pitch (10.0° at 30-inch radius, selected to give 2675 rpm at best climbing speed at 2,000 feet) and the level flight data with the propeller set in high pitch (12.75° at 30-inch radius, selected to give 2500 rpm at 2,000 feet in level flight with full throttle). The Phase IV tests were flown between May 4 and June 18, 1948, with additional tests in August 1949.

The potential gain in performance resulting from the installation of the Hartzell automatic two-position propeller on this airplane was 10 percent in takeoff distance, no change in climb performance, and a 6 percent increase in cruising speed with a corresponding 2 percent decrease in maximum speed. It was concluded the slight improvement in performance did not appear to justify the expense and complication of the automatic two-position propeller.[11] Numerical performance values for this test have not been presented in this book as they were flown, and the values given in the report for 2,448 pounds, which is 400 pounds or approximately 20 percent over the XL-15's gross weight.

During the Phase IV flight testing, the aircraft was flown to Fort Campbell, Kentucky, to take part in an Army demonstration. While there, it was caught in an unexpected windstorm and damaged. It was returned to Wright-Patterson in a C-82 transport airplane, which was a good demonstration of its air portability. The aircraft was repaired at Wright-Patterson AFB.[12]

On December 22, 1948, the second XL-15, 46–521, and the first YL-15, 47–423, flew from Boeing Wichita to Wright Field.[13]

YL-15 Performance

The YL-15 was 99 percent the same as the final configurations of the XL-15. It did, however, have a 50 pounds heavier gross weight. Since it had the same aerodynamic configuration, engine, and propeller, it was expected to have the same performance.

However, every Boeing and Air Force document the author has found has a different set of performance values. Table 6.5 presents that variety of performance in the YL-15 documents, and each will be discussed in the following paragraphs.

Table 6.5. YL-15 Performance

Requirement		Characteristics Summary	Pilot's Handbook	Pilot Comments	Pilot's Handbook	Standard Characteristics
Weight	lbs	2,100	2,100	2,100	2,100	2,100
TO dist over 50 ft—sod	ft	600	750	600	750	605
Landing dist over 50 ft—sod	ft	330	415	330	415	332
Landing speed	mph	31				
Max R/C	fpm	620	615	622	615	595
Service ceiling	ft	13,500	14,130	12,700	14,130	12,250
Cruise at 75% power	mph	94	95		91	
Range, with aux tank	s. miles	700		760		513
Endurance, standard fuel	hours	5.5		5.2		
Reference		WD-12280	AN 01–20LAA-1	WD12827	AN 01–20LAA-1	WD-12366
Date		27-May-48	1-Nov-48	10-Jan-49	23-Mar-49	15-Jul-49

In response to a letter dated May 6, 1948, to all U.S. aircraft companies from the USAF Air Materiel Command requesting characteristics data on their aircraft, Boeing L-15 Project Engineer E. O. Weining put together a characteristics summary for the YL-15. The takeoff, landing, and cruise speed numbers were referenced from AMC *Memorandum Report MCRFT-2120 (Addendum 1)*. The max rate of climb, service ceiling, range, and endurance were based on Boeing flight test results. It was also stated that the Boeing flight tests measured takeoff (608 feet) and landing (380 feet) distances that were greater than those used from the AMC report. The large difference in endurance was because the Boeing flight test was done at 65 mph while the Air Force test was done at 85 mph. Similarly, the Boeing maximum range was at 75 mph while the Air Force test was done at 81 mph.[14]

Performance for the YL-15 is given in the *Pilot's Handbook* dated November 1, 1948, which the author believes is a Boeing-written draft for the initial handbook. The maximum rate of climb and service ceiling were stated to be based on Boeing

Wichita flight tests. The cruise at 75 percent MCP, takeoff distance, and landing distance were stated to be based on AMC flight tests.[15] The *MCRFT-2120 (Addendum 1)* takeoff and landing tests were stated as minimum distance tests and the values in this handbook for takeoff and landing are exactly 25 percent higher than those in *MCRFT-2120 (Addendum 1)*. This increase in field distance may be a fudge factor added to account for average flying.

Doug Heimburger, Boeing L-15 Project Pilot, prepared a document to familiarize those pilots assigned to the YL-15 Service Test Program.[16] The takeoff and landing distances in this document were from *MCRFT-2120 (Addendum 1)*. The climb data were essentially the same as shown in other Boeing reports, but the hand fairing through the test points was slightly different. The maximum endurance and range numbers were determined using Air Force and Boeing flight test data, with the fuel consumption, increased 5 percent. The endurance was based on a maximum economy mixture setting flown at an indicated airspeed of 53 mph at 2000 ft. The maximum range was based on a maximum economy mixture setting flown at an indicated airspeed of 67 mph at 2000 ft.

The performance in the official YL-15 *Pilot's Handbook and Flight Operating Instructions* dated March 23, 1949,[17] is identical to that in the previous described handbook dated November 1, 1948, except for the 75 percent MCP cruise speed at sea level, which was 91 mph compared to the earlier version's value of 95 mph. The reason for this change in cruise speed is not known.

Boeing aerodynamicist James M. Wickham prepared a report containing the performance data and calculations for Standard Aircraft Characteristics Charts and Characteristics Summary Sheets for the Boeing YL-15 airplane. The performance was based on the testing described and data in *MCRFT-2120 (Addendum 1)* and in *WD-12321 YL-15 Cruise Control*. Takeoff, landing, and climb data were the results in *MCRFT-2120 (Addendum 1)* corrected for the weight increase from 2,090 pounds to 2,100 pounds. The range with an external tank was based on the flight test data in *WD-12321 Cruise Control Model YL-15* with the cruise portion flown at 5,000 feet with max economy mixture setting at 82 mph and included a 10 percent fuel reserve. The significant difference in range between *WD-12366* and the earlier *WD-12280* is because the earlier report did not have a reserve included and used a specific range of 16.8 statute miles per gallon of fuel, while this report had a reserve included and used a specific range of 13.4 statute miles per gallon of fuel based on the YL-15 flight tests in *WD-12321*.[18]

Earl Weining presented a paper[19] at the Seventh Annual Personal Aircraft Meeting of the Institute of Aeronautical Sciences on May 20, 1950, which included performance for the YL-15. Where the performance numbers came from was not stated in the paper, but they were the same as or better than those previously published. The range number was the exception and appears to be with only wing fuel (no external tank). This was the latest dated official document (Boeing or military) the author has found with performance data.

Looks at More Power

On August 15, 1946, less than a month after receiving the XL-15 contract, in a letter to Air Materiel Command, Boeing proposed a version of the XL-15 with a 140 hp Lycoming engine and designated Boeing Model 451–4a. The gross weight would go up 13 pounds.[20]

Table 6.6. Boeing Proposal—15 August 1946

Requirement		125 HP	140 HP
TO dist over 50 ft-sod	ft	595	530
Landing dist over 50 ft-sod	ft	517	517
Landing Speed	mph	35	35
Max R/C	fpm	650	800
Service Ceiling	ft	17000	19700
Cruise at 75% Power	mph	105	113
Endurance, standard fuel	hours	4	4

On July 7, 1947, less than a week before the first flight of the first prototype XL-15, Boeing sent a letter to Air Materiel Command recommending that a 170 hp Lycoming GO-290-A driving an 86-inch diameter Harzell propeller be installed in the second prototype, 46–521. Empty weight would increase to 1,626 pounds and gross weight to 2,166 pounds with the GO-290-A.[21]

Table 6.7. Boeing Proposal—7 July 1947

		O-290-7	GO-290-A
Requirement		125 HP	170 HP
TO dist over 50 ft-sod	ft	595	455
Landing dist over 50 ft-sod	ft	517	540
Landing Speed	mph	35	37.5
Max R/C	fpm	615	1040
Service Ceiling	ft	15,000	21,000
Cruise at 75% Power	mph	95	105
Range, with aux tank	s. miles	540	575
Endurance, standard fuel	hours	4	4

A conference on the YL-15 with Boeing, Army Ground Forces, and Air Force Air Materiel Command was held in Wichita on October 6–8, 1947. The primary reason for the conference was that the poor performance test results of the XL-15 indicated that the near-identical YL-15 would not meet its guaranteed performance or the minimum requirements of the Military Characteristics for Liaison Aircraft and would be of dubious value for tactical liaison operation. The YL-15, as specified, was considered too unsatisfactory by the Army Ground Forces to conduct functional and operational tests for evaluation as a combat article. The general opinion of the Army and Air Force was the airplane was underpowered. Replacing the 125 hp O-235-7 with higher power engines was a major discussion. Among the engines discussed were the Continental C-145, the Franklin O-335, and the Lycoming O-435. There was also discussion on different propellers and their advantages and disadvantages. The C-145 would require a firewall redesign, while the O-335 and O-435 would also need some airframe structural redesign due to weight increases. The estimated performance with the Continental and Franklin engines is presented in Table 6.8.[22]

6. Performance

140 HP Lycoming O-290-A2 engine in XL-15 sn 46–521 (NARA via AEHS).

Table 6.8. YL-15 Alternate Engine Proposals—October 1947

Engine		Lycoming O-290-7	Continental C-145	Franklin O-335	Franklin O-335
Horsepower	hp	125	145	150	150
Propeller		McCauley fixed pitch	Beech/Hartzell const speed	McCauley fixed pitch	Beech/Hartzell const speed
Gross Weight	lbs	2117	2160	2215	2215
TO dist over 50 ft-sod	ft	595	575	650	575
Landing dist over 50 ft-sod	ft	517	500	500	500
Landing Speed	mph	35	35	36	36
Max R/C	fpm	600	700		
Cruise at 75% Power	mph	90	99	99	99
Range, with aux tank	s. miles	500			
Endurance, standard fuel	hours	3.5	4	4	4

On December 5, 1947, the Air Force requested Boeing, at its own expense, to install a Continental C-145 engine with the most suitable propeller on the second XL-15, serial number 46–521, and flight test this installation. The Air Force pointed out that the purpose of this installation and test was not to get improved performance over the existing XL-15 specification, but simply to get the contractually guaranteed performance. If the C-145 equipped XL-15 met the guaranteed performance, the government would be willing to renegotiate the YL-15 contract to use the C-145 in place of the O-290-7 engine. The author did not find Boeing's reply to this, but since the C-145 engine installation did not occur and the YL-15s all used the O-290-7 engine, it is assumed that Boeing turned down this request.[23]

In a conference of Army and Air Force personnel on November 3, 1948, it was decided that the Army Field Forces (the new name for Army Ground Forces) would accept the ten O-290-7 powered YL-15s for service testing and tactical evaluation. This decision was made with the understanding that a test program using a higher horsepower engine would be run to determine what the increase in performance would be. Since Boeing would not do such a program without additional funding and the Army Field Forces had no funds available for such a program, Air Materiel Command agreed it would do the program in-house using the second XL-15 and an O-290-A2 engine loaned at no cost by Lycoming.[24]

XL-15 sn 46-521 was reequipped with a 140 hp Lycoming O-290-A2 engine with a McCauley 1A170-L8046 propeller modified to a 13.75° pitch at 0.75 radii for the test. Standard YL-15 sn 47-423 was flown as a comparison. The same crew, pilot USAF Captain Thomas J. Ceicl of the Air Materiel Command Cargo, Training, and Miscellaneous Aircraft Section and flight test observer Jack H. King, an engineer in the Performance Engineering Section of the AMC Test Engineering Section, alternated flights in the two aircraft to get the fairest comparison. Both aircraft had a test weight with test instrumentation, crew, and fuel of 2,230 pounds at 44.4 percent MAC. This testing was done at Wright-Patterson Air Force Base from June 22 to August 3, 1949, and consisted of 9 flights totaling 15 hours and 20 minutes flying time. Test results are presented in Table 6.9 for the two airplanes, plus the test results from the official XL-15 tests corrected to the same 2,230-pound weight.[25]

Table 6.9. Increased Power Performance

Requirement		USAF XL-15 Flight Tests	YL-15	XL-15 with 140 HP Engine
Power	HP	125	125	140
Weight	lbs	2,230	2,230	2,230
TO dist over 50 ft-sod	ft	685	805	980
Max R/C	fpm	545	475	555
Service Ceiling	ft	10,000	10,600	11,850
Cruise at 75% Power	mph	94		89
Reference		MCRFTP-2120 (Addendum 1)	MCRFT-2243	MCRFT-2243
Date		6-Feb-48	1-Sep-49	1-Sep-49

The purpose of this test was to look at the higher horsepower, but the difference between that of the YL-15 compared to the earlier XL-15 test, which should have been zero, really stood out. To quote the report, "This discrepancy ... cannot be satisfactorily explained. It is possible that errors in measuring the earlier takeoffs or possible deficiencies in the data reduction methods as applied to small, slow speed, aircraft of this type, may account for some of the difference. In any case, the data included in this report is believed to be most representative of the performance of the YL-15 aircraft."[26]

The report also stated that "the power difference, indicated by the increase in performance, was approximately seven thrust horsepower at sea level."[27]

Even with the 140 horsepower engine, the XL-15 tests indicated the aircraft could not meet contract guarantees.

7

Airframe

The Boeing L-15 was unusual for an Army liaison airplane of its day as it was an all-metal aircraft with a cantilever wing and a twin tail. The previous liaison airplanes had primarily been welded steel tube fuselage fabric-covered airplanes with strut-braced wings. Except for the L-13 that was developed for the Air Force, not the Army, all previous liaison aircraft had been designed to Civil Aeronautics Administration (CAA) requirements. Boeing designed the L-15 structurally to AAF Specification C-1803-D, except the load distribution and maneuvering loads, which were done to Civil Air Regulations Part 03.[1]

For stress analysis purposes, the design gross weight was 2,050 pounds, and the design limit maneuver and gust load factors were positive 4.0 and negative 2.0. Limit diving speed was 200 mph IAS.[2]

The Project Stress Engineer was Wilfred (Bill) A. Pearce, a 1937 BS in Aeronautical Engineering graduate of the University of Oklahoma. Bill went to work for Boeing Wichita (then Stearman) when he graduated. On the L-15 project, Bill led a group of many stress engineers.

Since this was a Boeing design, it had a very detailed structural analysis that consisted of hundreds of pages in many different engineering reports. It was ten or more times greater in size than the analysis done by the light aircraft companies of that era.

Wing

The wing assembly consisted of left and right-hand wing panels, each attached to the top of the fuselage by four special close-fitting hinge bolts through wing root wing to body 14ST forging attachment fittings riveted to the spar caps. The wing attachment forgings had replaceable X4130 flanged-type steel bushings at the attachment points. The wing hinge bolts were designed to facilitate assembly and disassembly. The lower two wing attachment fittings incorporate a hook that engages a shoulder bushing pressed into the fuselage fitting to aid attachment of the wing panel to the fuselage. The body gap between the wing and fuselage was covered by a wrap-around sheet metal cover fastened by one AN machine screw at the trailing edge.

Each wing panel was a full cantilever semi-monocoque two-spar, stressed skin structure made of aluminum alloys. The spar caps were 24ST T-shaped extrusions

7. Airframe

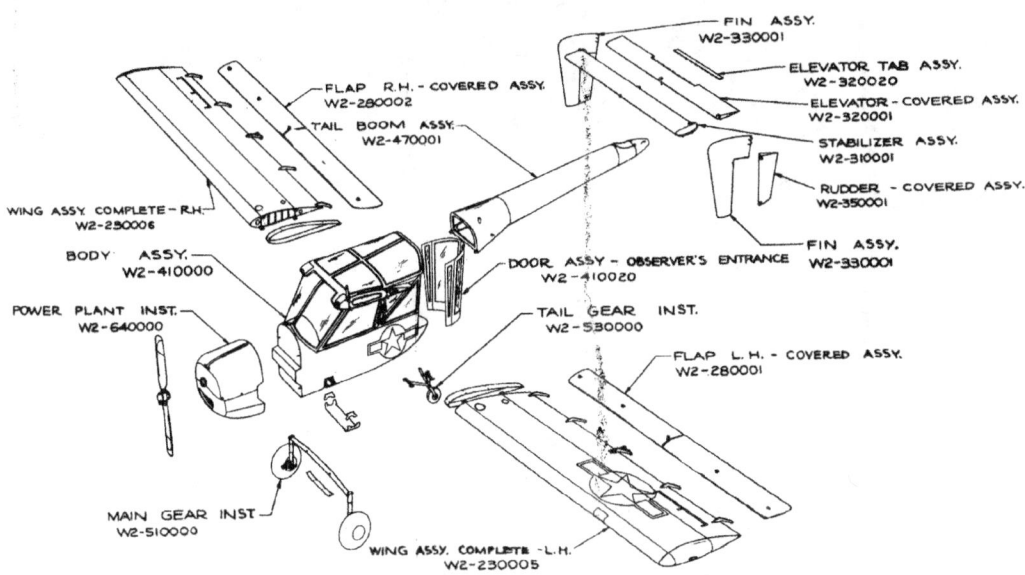

Exploded view—major components of the XL-15 airplane (NARA).

with an additional 75ST cap strip on the top flanges and a 24ST strip on the bottom flanges, added where the extrusion by itself became critical and tapered in thickness to obtain the required area at the inboard stations. Spar web stiffeners were formed angles of 24ST aluminum alloy.

The front spar web thickness was 0.032 inches from the root (WS 26.9) to WS 50, 0.025 inches from WS 50 to WS 76, and 0.020 inches from WS 76 to WS 235.5. The rear spar web thickness was 0.032 inches from the root (WS 26.9) to WS 50, 0.025 inches from WS 50 to WS 110, and 0.020 inches from WS 110 to WS 235.5.

Wing torsion was taken out in the torsion box between the two spars formed by the spar webs and wing skin, both made of 24ST aluminum alloy. The section of the wing between the spars was made up of 24ST aluminum alloy main ribs spaced every 48 inches, which also were the flap hinge stations, and 24ST aluminum alloy formers spaced eight on top and five on the bottom between the main ribs. The formers were supported by vertical stiffeners at points that were found critical in the wing structural testing. The inboard bay of each wing was a fuel cell compartment with access doors above and below.

The leading edge wing skin thicknesses were 0.020 inches, and the trailing edge skin thicknesses were 0.016 inches. The wing skin thickness between the spars was 0.025 inches from root to WS 75.7, 0.020 inches from WS 75.7 to WS 171.7, and 0.016 inches from WS 171.7 to the tip.

There were five small hinged doors on the lower surface of each wing panel to provide access to electrical and control cables. The left-hand wing leading edge had a cutout for twin landing lights with a transparent cover.

The exterior airfoil full-span flap was mounted from the trailing edge upper surface of the main wing ribs by 24ST aluminum alloy hinge fittings.

Wing rear spar and ribs on buildup jig (author's collection).

Wing structural testing was performed by the Air Materiel Command at Wright Field. Five loading conditions were tested, and three of them carried 100 percent of the ultimate load. One test failed at 90 percent of ultimate, requiring beef-ups by the addition of stiffeners. The fifth test failed at 98.5 percent on the left wing, which had failed in one of the previous tests, but went to 100 percent on the right wing and was considered good.[3]

Flap

The flap was a full-span external airfoil type with the hinge located at the 25 percent flap chord to give low operating loads. It operated as both a high lift device and asymmetrically for lateral control as an aileron.

The flap construction was a metal frame with D-nose and fabric covering. It had a constant cross section built in two panels (per side) 107.5 inches long each. A sheet metal gap closure cover covered the gap between the two panels. The inboard half of one flap was interchangeable with the outboard half of the other flap. Each flap was supported by five hinge brackets and was divided into two parts with a common control pull-off at the center hinge bracket. The flap spar was constructed

Wing showing left spoiler fully up and full span external airfoil flap (author's collection).

of 2.84 inches wide 0.032-inch 24ST aluminum with 1.8-inch-diameter lightening holes. All flap nose and trailing edge ribs were "C" section type constructed of 24ST with a thickness of 0.016 inches except for the nose ribs at stations 123.57 and 124.27, which were 0.051 inches. The leading edge D-nose skin was 0.016 inches thick from WS 21.90 to WS 66.3 and WS 181.5 to WS 225.90 and was 0.020 inches thick from WS 66.3 to WS 181.5.

The flap loads were based on the airplane condition of flaps down 40° at 80 mph with a positive gust. This load exceeded the required loads and was therefore conservative. Static tests for this condition were done by the Air Materiel Command.[4]

Spoiler Ailerons

The spoiler aileron was a circular-arc type located between wing stations 172 and 220 and was mounted on three hinges in a recess in the upper wing surface aft of the rear spar. The concavity of the circular-arc spoiler faced forward. The blade of the spoiler was formed of a 0.076-inch-thick magnesium sheet and was attached to three aluminum alloy ribs that contained the hinge bearings. The circular arc spoiler had a radius of 4.98 inches and stuck up approximately 4 inches above the wing surface when fully deployed. When retracted, the spoilers lay in the recesses flush with the upper wing surface. The control actuation was at the center hinge. Left and right spoilers were interchangeable.

Circular-arc spoiler aileron (author's collection).

The Air Materiel Command performed static tests on the spoilers to simulate air loads in a 200 mph dive.[5]

Horizontal Stabilizer

The horizontal stabilizer was a one-piece rectangular planform with two-spar metal construction using 24ST aluminum alloy, including the covering. The front spar was a stamped pan type, while the rear spar was built up of extruded caps and sheet web. Ribs were formed from sheet stock with flanged holes for stiffness. Chordwise beaded skin was used to eliminate intermediate ribs. It was attached to the tail boom using three AN bolts through two fittings on the rear spar and one extended forward from the station zero rib. There were three hinges mounted off the rear spar for the elevator.

The Air Materiel Command did static structural load testing for balancing tail loads, maneuvering down load, tab load, and unsymmetrical combined tail load.[6]

Elevator and Tab

The single-piece elevator was a fabric-covered aluminum frame hinged to the stabilizer at three points. Torsion was carried by a D-nose box. The elevator had no aerodynamic balance but was 100 percent statically balanced by using a weighted arm attached to the elevator control bell crank. On the centerline of the upper stabilizer surface, an access door provided access to the rudder, elevator, and elevator tab control mechanisms.

There was a partial-span trim tab inset into the center of the elevator trailing edge. The tab had a channel spar and aluminum alloy skin and was hinged off the elevator auxiliary spar with seven short lengths of piano hinge.[7]

Vertical Fin

The L-15 has two vertical fins mounted on each end of the horizontal stabilizer. They were unusual in that they were an inverted type, mounted downward instead of extending upward from the stabilizer in the conventional manner. The fin was of all-metal construction, with a pan-type front spar and a built-up rear spar of 24ST sheet and extrusions covered with a chordwise beaded skin. Each fin was attached to the outboard end of the horizontal stabilizer by a single fitting on the front spar and two fittings on the rear spar using three AN bolts. The lower attachment bolts acted as hinges so the fin and rudder assembly could be folded inboard against the horizontal stabilizer to facilitate truck or aircraft loading. The right and left fins were interchangeable.

Static structural load testing was done on the fin and rudder by the Air Materiel Command for ground gust load, steady yaw load, and a modified steady yaw load.[8]

Rudder

The rudder was of conventional D-nose construction with a single spar and aluminum alloy ribs and formers. It was fabric covering. Two hinges attached the rudder to the fin.

Each rudder had a horn balance that provided some aerodynamic balance. The rudders were also 100 percent statically balanced. The two rudders were interchangeable except for a small ground adjustable metal tab attached by two bolts and nuts to the left-hand rudder.[9]

Fuselage

The L-15 has a fuselage made up of a gondola-type cabin and a tail boom. The cabin was made up of two separate compartments. The forward compartment, or pilot's cockpit, extends from the firewall aft to fuselage station (FS) 111.45 and contains the pilot's seat, the instrument and switch panels, flight, engine, and communication equipment controls. Directly behind the pilot's seat were located the control cables, which passed vertically from the floor level to the top of the fuselage and led to the empennage and wings. The control cables were covered by a latched-on sheet metal cable guard and constituted the separation between

Pilot's cockpit (author's collection).

compartments. Aft of the cable guard, the aft, or observer's, compartment extended to FS 162, where it was closed in by a pair of transparent plastic doors. This compartment contained a swiveling observer's seat; also, rudder pedals, a control stick, and a throttle control which were all connected to corresponding controls in the cockpit.

The lower portion of the fuselage cabin up to the window/door line was a semi-monocoque, aluminum alloy, riveted structure. The fuselage sides consisted of skin paneling approximately 18 inches up from the floor line, with transparent panels, windows, and doors to the top. The top of the fuselage was constructed with a front and rear spar (these being extensions of the wing spars) with aluminum alloy longerons, frames, and braces. The fuselage top, between the front spar and the tail boom, also consists of transparent sheet plastic, and the windshield panels were unusually wide and high.

There were separate doors for the pilot and observer. The cockpit door was on the right-hand side of the airplane with one sliding and one fixed transparent plastic window panel in an aluminum alloy frame and was hinged to swing out and aft. Its purpose was to provide regular and emergency access to and exit from the airplane for the pilot. A door latch was on the forward edge with yellow knobs inside and out. When the door was closed, the pilot must lock the latch in position by pushing

Left: Observer's compartment (author's collection). *Right:* Bottom section of gondola without skin (author's collection).

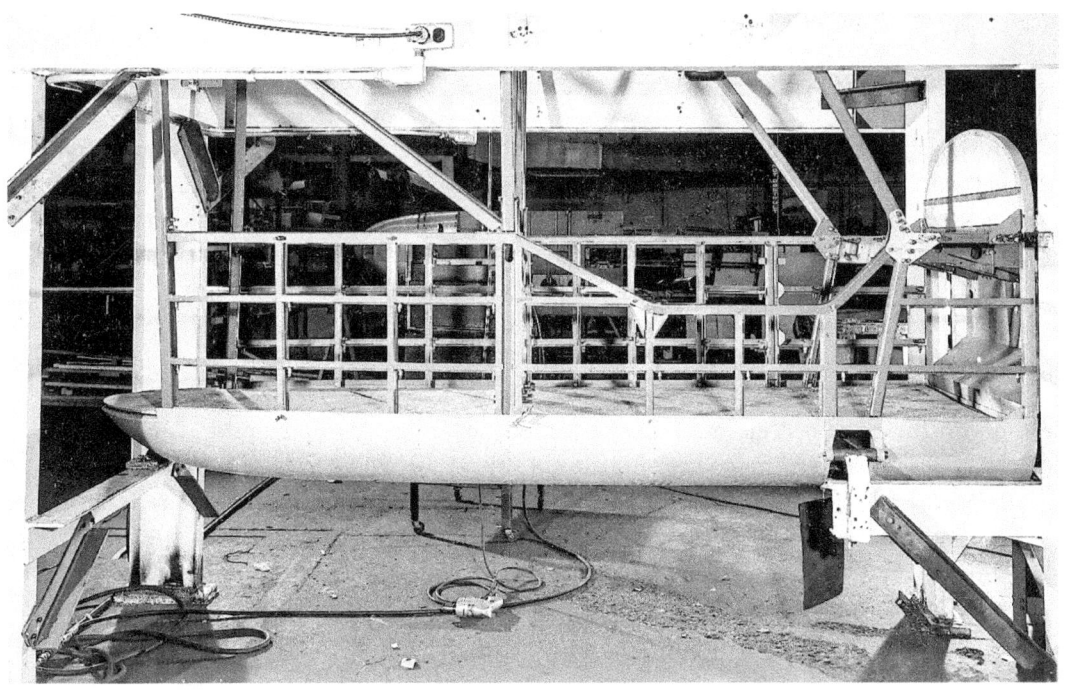

Fuselage in construction jig (author's collection).

Fully open pilot's door on XL-15 on floats (author's collection).

downward on the yellow handle inside the pilot's compartment. There were hand grips on the lower edge of the door, both inside and outside the airplane. When fully opened, the door rested against the side of the observer's compartment and held open with a catch, set in the fuselage. The catch was accessible from the observer's compartment.

Before the prototype was built, Boeing engineering used the mockup to evaluate different hinge locations for the pilot's door, such as at the rear of the door, front of the door, and top of the door. They chose to go with the hinges at the rear of the door. During the accelerated service test of the YL-15, Air Force pilots several times had difficulties closing or keeping closed this aft hinged door when the engine was running. The accelerated service test report recommended the door be redesigned. The author has found no record of where this was done. All photos of the YL-15 in military and Fish and Wildlife service show the aft-hinged suicide door configuration. However, the restored YL-15 47–432 (N4770C) has the pilot's door hinged on the top, opening up against the wing. It is not known when this modification was made.

An emergency exit from inside the cockpit was accomplished by pulling down on the cockpit door emergency release handle located inside the door near the upper aft corner. This handle was a yellow-colored ring grip at the end of a cable and was

Pilot's emergency door release (author's collection).

located above the upper aft corner of the door. When you pulled the handle, the entire cockpit door would drop off.

In a ground emergency, quick access to the cockpit from outside could be gained by turning two quick-release Wedjit fasteners on the outside door hinge line. The handles of these fasteners were painted yellow. Turning them in either direction allowed the door to drop off its hinges and fall outward.

A cockpit window was installed on the left-hand side of the fuselage just aft of the left-hand windshield panel. It consists of a single transparent plastic window panel supported in an aluminum alloy frame hinged on the lower edge, opening outward and down. A handgrip on the inside upper edge was provided for use in opening and closing the window. A latch was provided to hold the window in the open position.

There were three windows on each side of the observer's compartment. Each consisted of a triangular, transparent plastic panel in an aluminum alloy frame. The upper triangular windows were hinged on their upper edge and swung out and up with a latch and knob handle at the lower forward corner. There was a catch to secure them in the open position.

Access to the observer's compartment was through the rear of the fuselage,

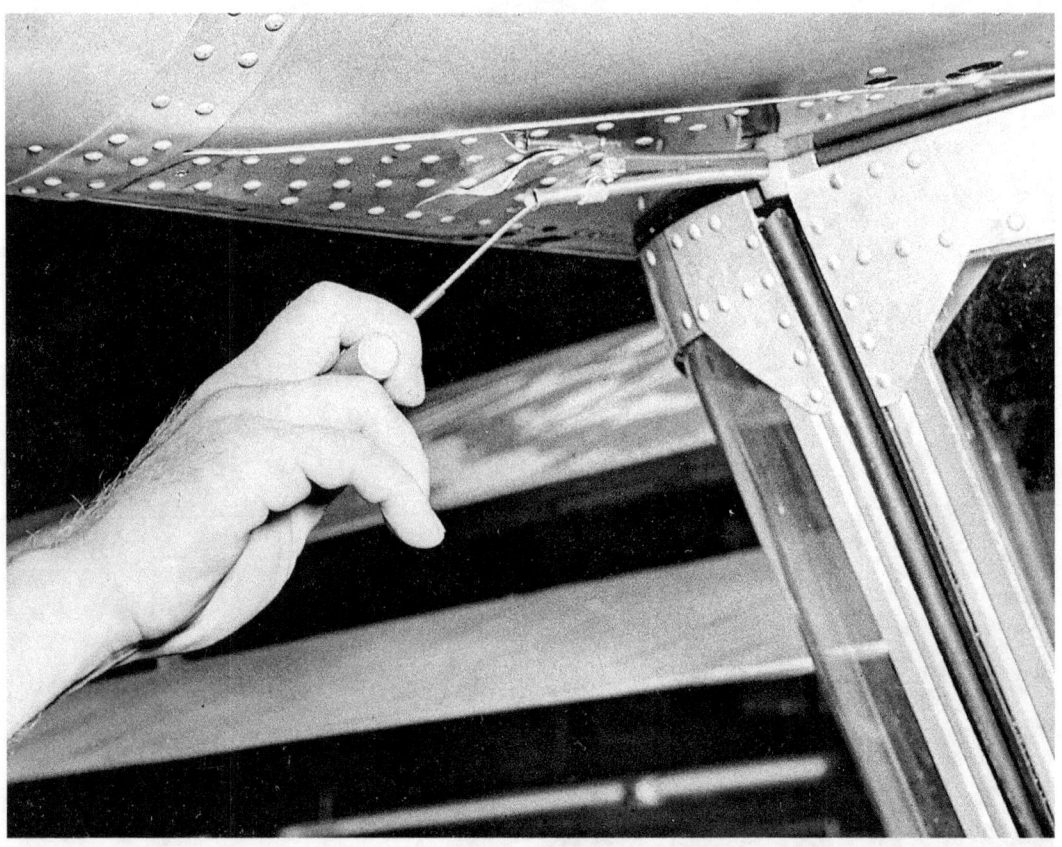

Exterior observer's door emergency release (author's collection).

which was closed by a pair of full-length vertical doors. Each door consisted of a full-view transparent plastic panel mounted in an aluminum alloy frame, and the two were hinged to swing sidewards. The observer's doors opened and closed in unison and were secured in the closed position by a latch midway up on the edge of the left door with yellow color tee handles inside and out. The doors were supported in aluminum alloy frames and operated together with an assembly of connecting rods and springs, located in the upper part of the triangular panels (called observer's compartment door supports).

There was also a jettison-type emergency observer's door release which had a handle both inside and outside the airplane. Emergency exit from inside the observer's compartment was accomplished by pulling down on the door emergency release handle. This was a yellow-colored grip on the release cable and was held in two clips on the frame, directly above the observer's seat to the left of the data case. The doors would fall free of the airplane when this grip was pulled. In the event of a ground emergency, quick access to the observer's compartment could be gained from outside by pulling down on the exterior handle of the observer's compartment door emergency release. This was a yellow-colored tee handle on the release cable that was secured in a clip outside the doors beneath the fuselage boom. The doors would fall free of the airplane when this handle was pulled.

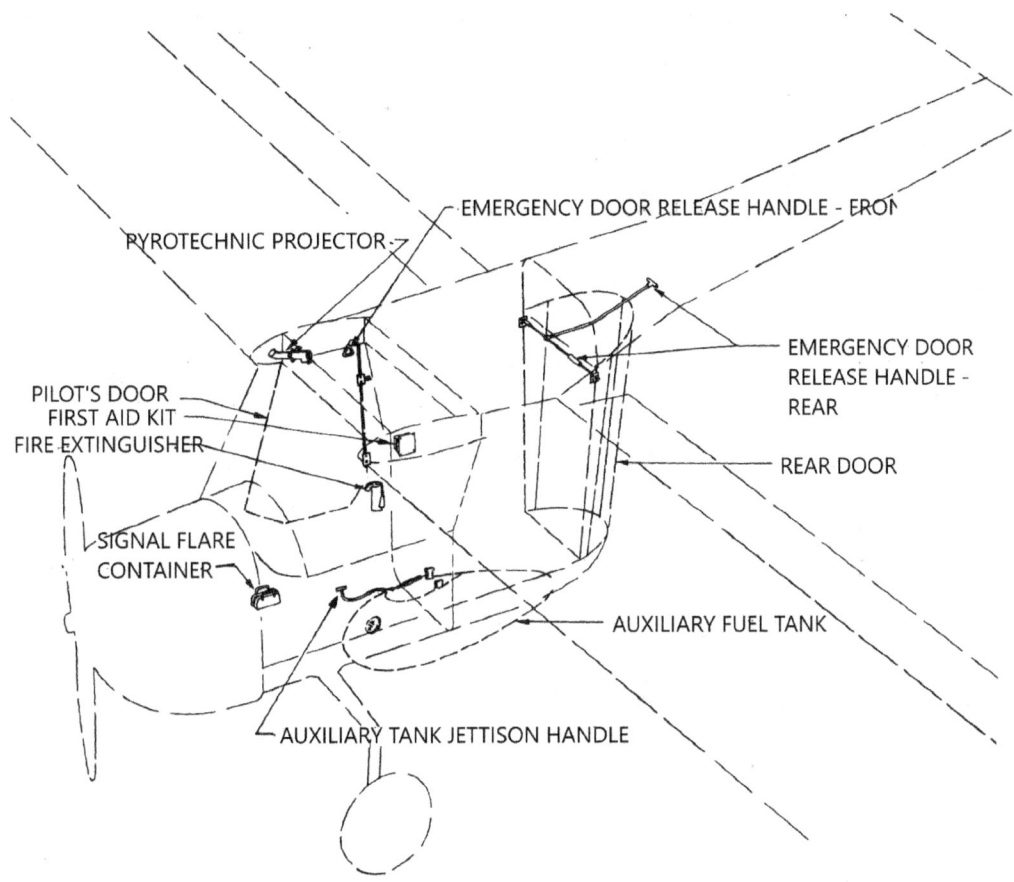

XL-15 emergency equipment and exits (NARA).

The observer's doors could be removed with the doors closed and latched by pulling down on either handle of the observer's compartment door emergency release, making certain that the doors and support assembly were held in place when the release handle was pulled. If not supported, they would fall aft and downward. The doors and support assembly could then be lifted up and off the left- and right-hand lugs in the floor for removal.

Thus the observer's doors provided both normal and emergency access to and exit from the observer's compartment. They also formed the fuselage aft fairing between floor level and the lower surface of the tail boom.

There were two assist handles or grips, provided for use in entering or leaving the airplane. The pilot's assist handle was bolted to the lower flange of the forward center section spar in the upper-right corner of the cockpit. The observer's assist handle was attached to the frame on the left forward corner of the observer's compartment.

Aft of the observer's compartment, the fuselage was extended in the form of a tail boom, with the empennage structure attached to its aft end. The tail boom consisted of a semi-monocoque, aluminum alloy tubular structure with transverse

Tail boom in construction jig (author's collection).

former rings and longitudinal stiffeners. The tail boom was supported by a triangular steel tube truss bolted to the top of the rear spar and to the upper rear vertical members of the cabin superstructure. Three quick attach fasteners were used in connecting the tail boom to the gondola.[10]

Fuselage Equipment

Parking Harness

The parking harness consisted of a scissor clamp with two cables and two web straps attached. Its purpose was to lock the control surfaces when the airplane was parked. The clamp was placed over the pilot's control stick just below the grip. Next, the strap ends were hooked to the left and right observer's rudder pedals, then the cables fastened to the left and right keyhole fittings on the dash below the instrument panel, and cables and straps were tightened by closing the lever lock on the clamp initially placed around the control stick. The adjustable web straps could be tightened should the harness assembly get out of adjustment. When not in use, the parking harness was placed in a stowage bag to the left of the pilot's seat.[11]

Seats

A conventional aircraft fixed seat with a cushion and headrest was installed in the pilot's cockpit. The pilot's seat was designed for a 40 g crash load. A safety lap belt and shoulder strap assembly were provided for the pilot. The shoulder straps were attached to an inertia reel at the rear of the seat. Tension on the reel was adjusted by a conventional lever lock on the cockpit floor at the left front corner of the pilot's seat. The shoulder harness could be locked by depressing the lever and moving it to the full forward position. To unlock, the lever was depressed and moved to the full aft position. The shoulder harness lock mechanism was constructed so that, should the pilot fail to lock the harness, it would automatically lock with the application of a 2 to 3 g inertia load on the shoulder harness. The harness would stay in the locked position until the normal locking lever was moved forward to the locked position and then returned aft to the unlocked position.

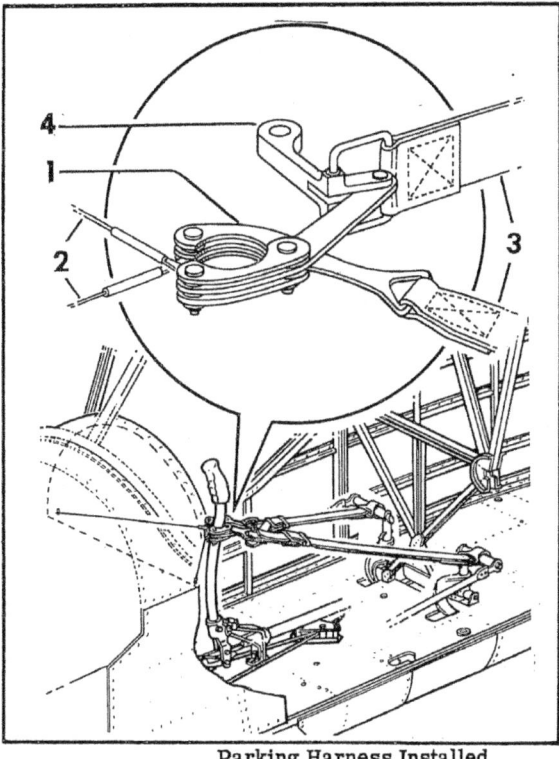

Parking Harness Installed

1. Clamp
2. Cables
3. Straps
4. Clamp lever

Parking harness for locking the controls (author's collection).

A swiveling seat, cushion, and headrest were installed in the observer's compartment. The observer's compartment seat had a lever lock to secure the seat in position facing forward or aft as desired. The seat could only be rotated when not occupied. There was a lap belt for the observer.[12]

Relief Tubes

There was a conventional horn-type relief tube in the cockpit and another in the observer's compartment. Both were connected by flexible hoses, joined at a tee fitting, to a venturi discharge vent. The vent was located on the lower surface of the fuselage just forward of the tail wheel.[13]

Flight Report and Map Case

An aluminum flight report and map case with a canvas flap cover was attached to the fuselage structure above the pilot's head. Snap fasteners held

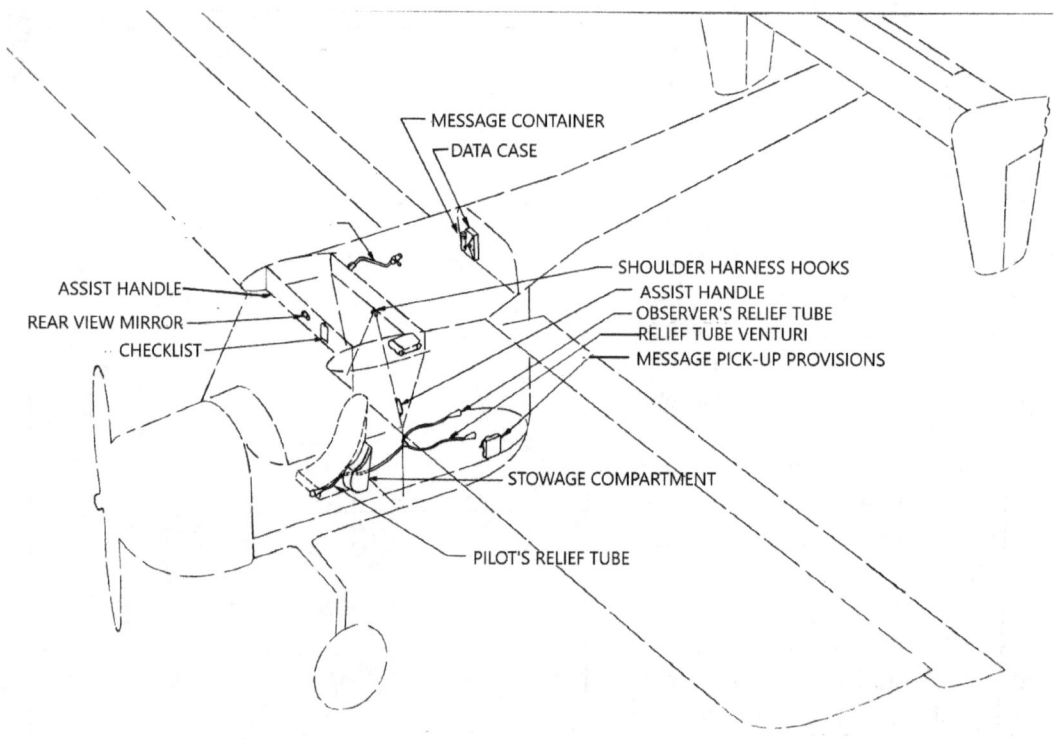

XL-15 miscellaneous equipment (NARA).

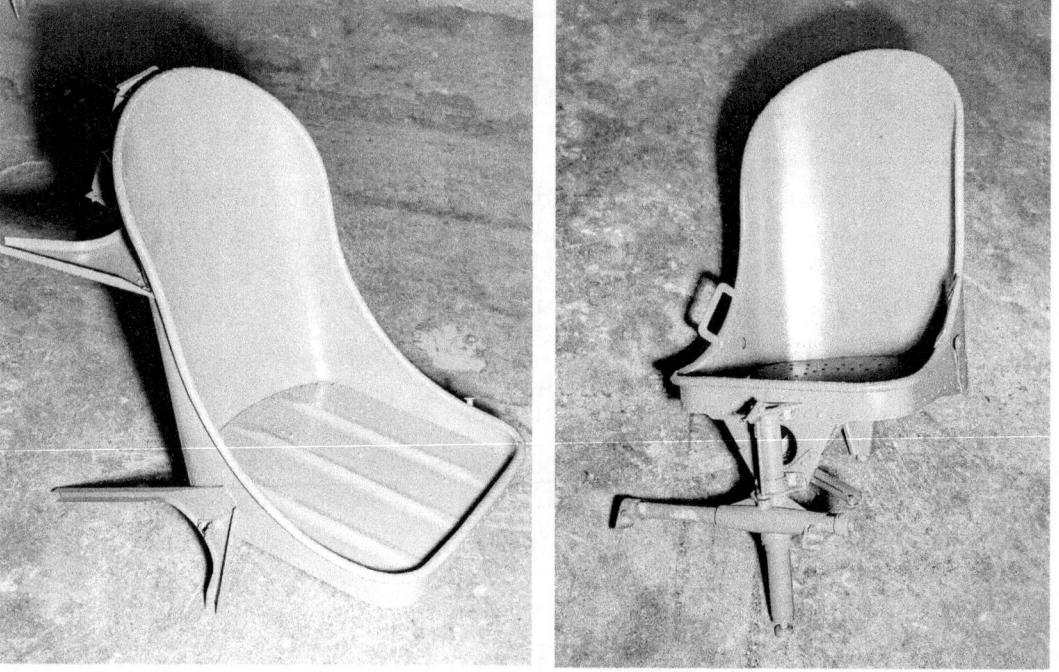

Left: Pilot's seat (author's collection). *Right:* Rotatable observer's seat (author's collection).

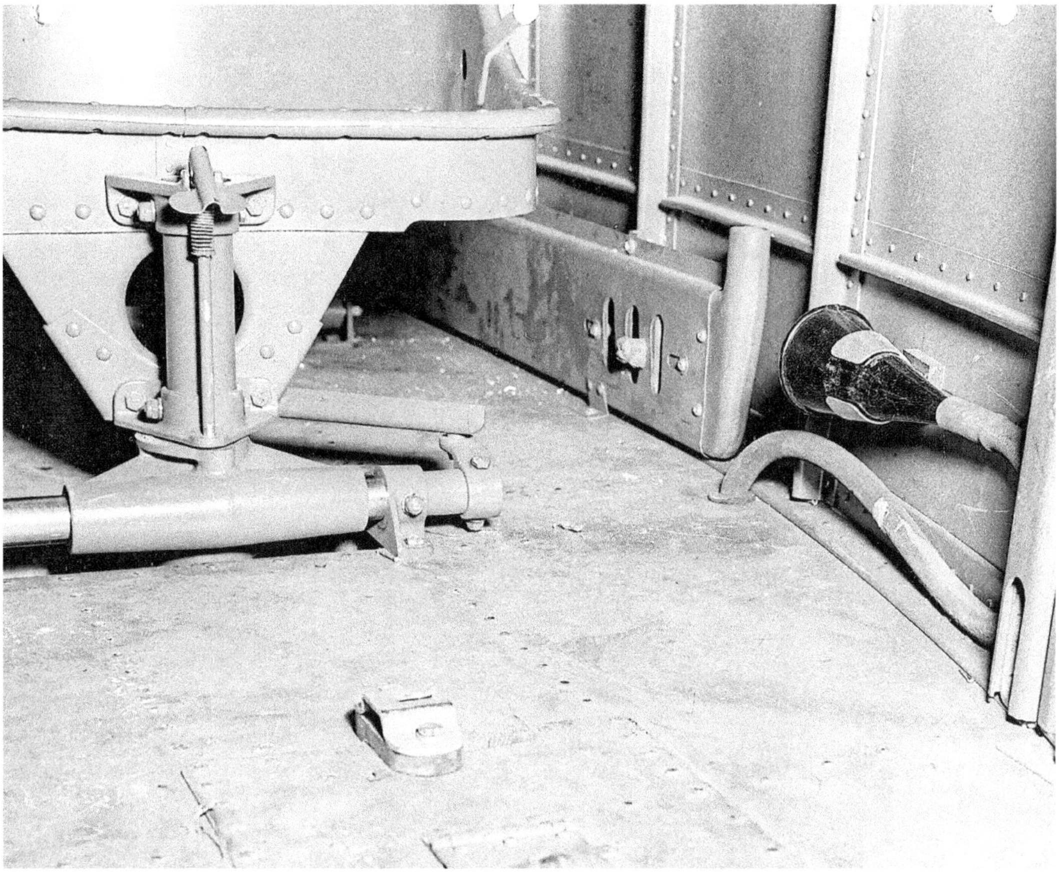

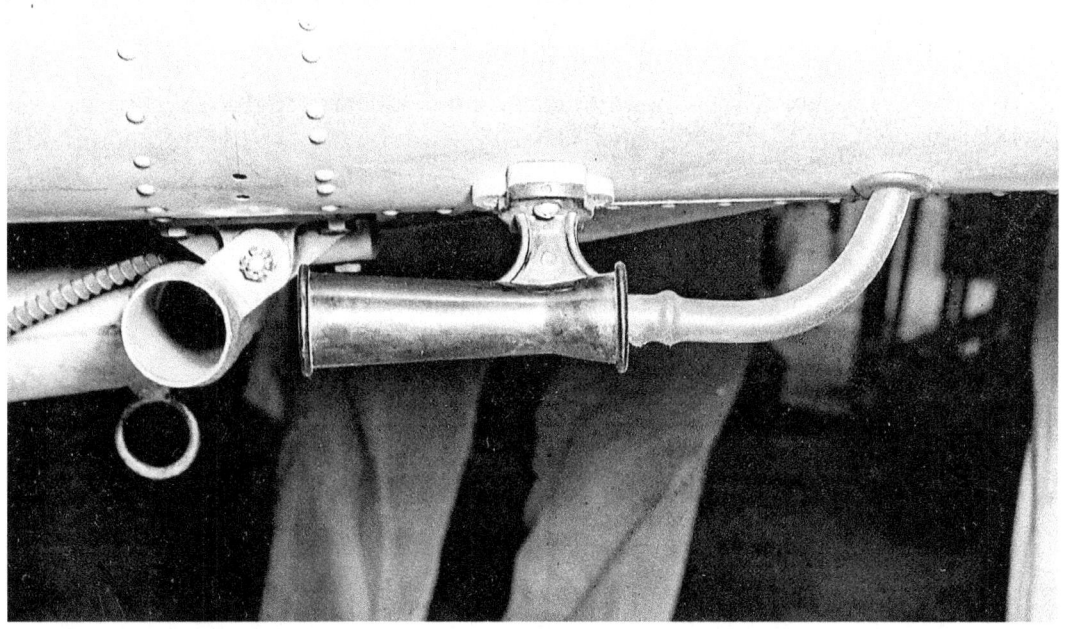

Top: Observer's relief tube (author's collection). *Bottom:* Relief tube venturi discharge vent (author's collection).

Flight report/map case and first aid kit (author's collection).

the flap cover closed. It was used to contain maps, charts, flight reports, etc., as necessary.[14]

First-Aid Kit

A first-aid kit was installed directly in front of the flight report and map case above the left-side cockpit window. It was secured with snap fasteners.[15]

Data Case

An aluminum data case with a hinged cover was located in the observer's compartment over the doors. The data case was used to stow the airplane handbooks. A small latch held the hinged cover in place. Clips for holding two message containers were attached to the data case.[16]

Message Pick-Up Equipment

Provisions were made for the installation of message pick-up equipment. When not in use, provisions were made for the stowage of message pick-up equipment to the left of the observer's compartment.[17]

Data case to stow the airplane handbooks (author's collection).

Message Containers and Bags

Two type A-8 message containers could be mounted on pairs of clips attached to the bottom of the data case located aft and above the observer's seat. A type A-1 drop message bag was attached to the left-hand side of the observer's compartment just above the floor level. Additional message container bags were stowed in a chamber beneath the floor of the observer's compartment just forward of the observer's seat. The door to this compartment was hinged at the forward end and was secured at the aft end by a push-to-open type quick release.[18]

Rear Vision Mirror

A rearview mirror was mounted on the aft face of the front center section spar forward of the pilot's seat. It was in a position that allowed the pilot a view toward the rear of the airplane.[19] However, the author suspects the control cable guard would have been a real blind spot in the mirror view.

Fire Extinguisher

A carbon dioxide hand fire extinguisher, type 2TB, was installed in the cockpit at the right-hand side of the pilot's seat. The pistol grip was accessible to either pilot or observer.[20]

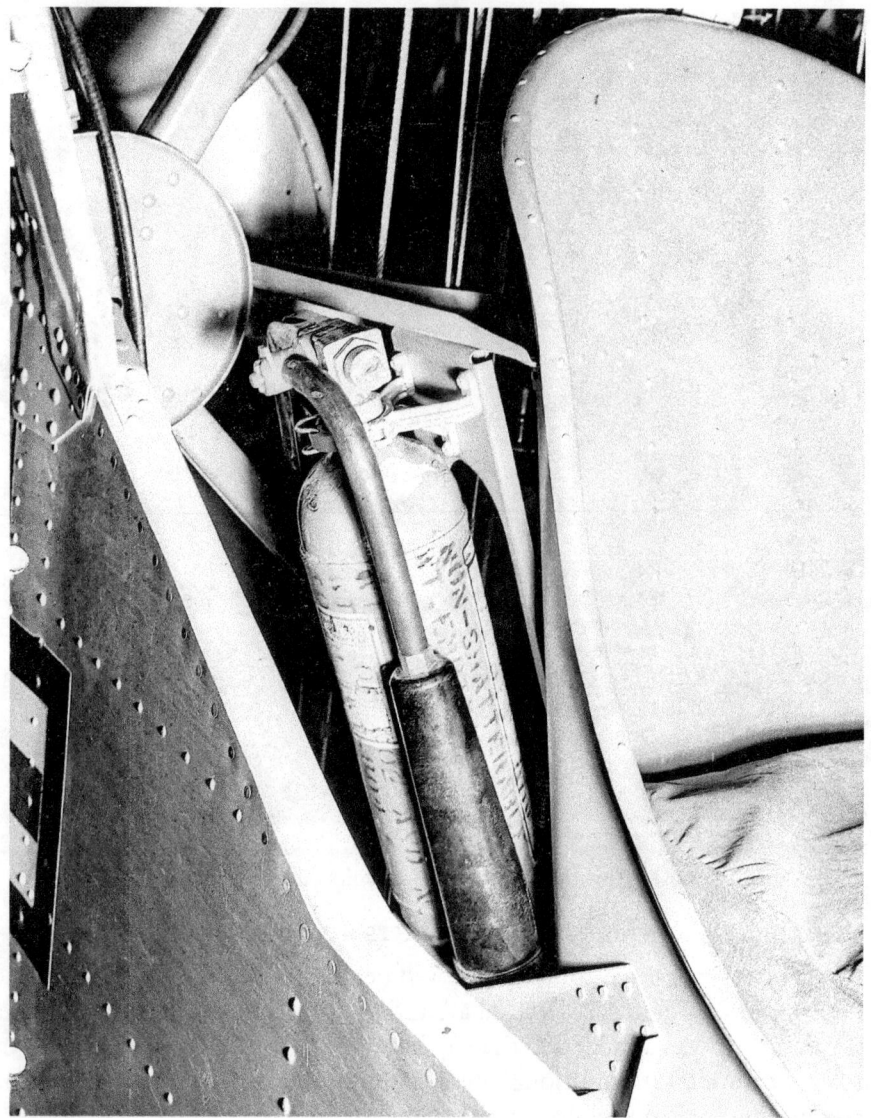

Fire extinguisher on the right-hand side of the pilot's seat (author's collection).

Pyrotechnic Projector

A type M-9 hand pyrotechnic projector (flare gun) was installed on a bracket in the upper-right-hand corner of the cockpit above the cockpit front window. The barrel projected through the wing leading edge and shot the signal flares upward and forward at approximately 45°. The breech-loading and trigger mechanism was accessible from inside the cockpit, and the whole unit could be removed by taking out four attachment screws that secured the projector to its bracket. You swung the breech down and open to load the projector, inserted the signal flare, and closed the breech. Pull the lock pin, fastened to the end of a short chain, from the firing pin. Fire the flare gun by firmly striking the firing pin with your hand.[21]

Flare gun in the upper right-hand corner of the pilot's cockpit (author's collection).

Signal Flare Container

Six aircraft signal flares were normally stowed in a type A-5 signal flare container located on the battery access cover, just aft of the pilot's rudder pedals. The container was a canvas bag, closed with an interlocking slide fastener and attached to the cover with snap fasteners.[22]

8

Propulsion

Engine

The power plant for the L-15 was a Lycoming O-290-7 direct-drive four-cylinder opposed air-cooled engine manufactured by the Lycoming division of the Aviation Corporation (AVCO) and described in Lycoming Specification No. 21028-A dated 9 August 1945. The O-290-7 had a 4.875-inch bore and a 3.875-inch stroke with a 6.5:1 compression ratio. Maximum normal power 125 bhp at 2600 rpm at sea level ISA. Takeoff power was 130 bhp at 2800 rpm at sea level ISA. Idle speed was approximately 550 rpm. The uninstalled engine had a dry weight of 259.4 pounds. The prescribed fuel for the XL-15 was 73 octane gasoline, while that for the YL-15, depending on the document, was either 73 or 80 octane.[1]

The engine mount was an X4130 seamless steel tube assembly and was attached to the fuselage by four AN steel bolts passing through conical rubber vibration absorber bushings on each side of the engine attachment lugs.

The engine cowling was supported from the firewall and completely isolated from the engine. The upper cowling consisted of two aluminum panels hinged at the top centerline and fastened down by Dzus fasteners. There were folding supports to hold the upper cowl doors in the open position for engine maintenance. The nose bug was formed from aluminum alloy. The aluminum lower left, center, and right cowling panels were held in place with flush head screws set in nut plates and were readily removable if needed to access the engine or accessories. The cowling was provided with chafing strips where required. The lower center fixed panel had an adjustable engine cowl for adjusting air circulation in the engine compartment. The engine oil filter was accessible through a door on the right-hand side. The cowling was designed to allow for easy removal of rocker box covers.

The power plant installation was designed so that the engine could be removed and replaced to operating condition in 45 minutes or less with four men.

The engine control system consisted of an Army Air Forces type B-17 engine control unit in the front pilot's cockpit that had a throttle lever with a microphone switch and a lever with a knob for carburetor mixture control and a type B-0 throttle control unit in the rear observer's cockpit that had a throttle lever with a throttle control quick release. Both engine control quadrants were on the left side. A control rod connected the two throttles so they could operate in unison. The throttle control quick release allowed the observer's control to be disengaged from the pilot's control

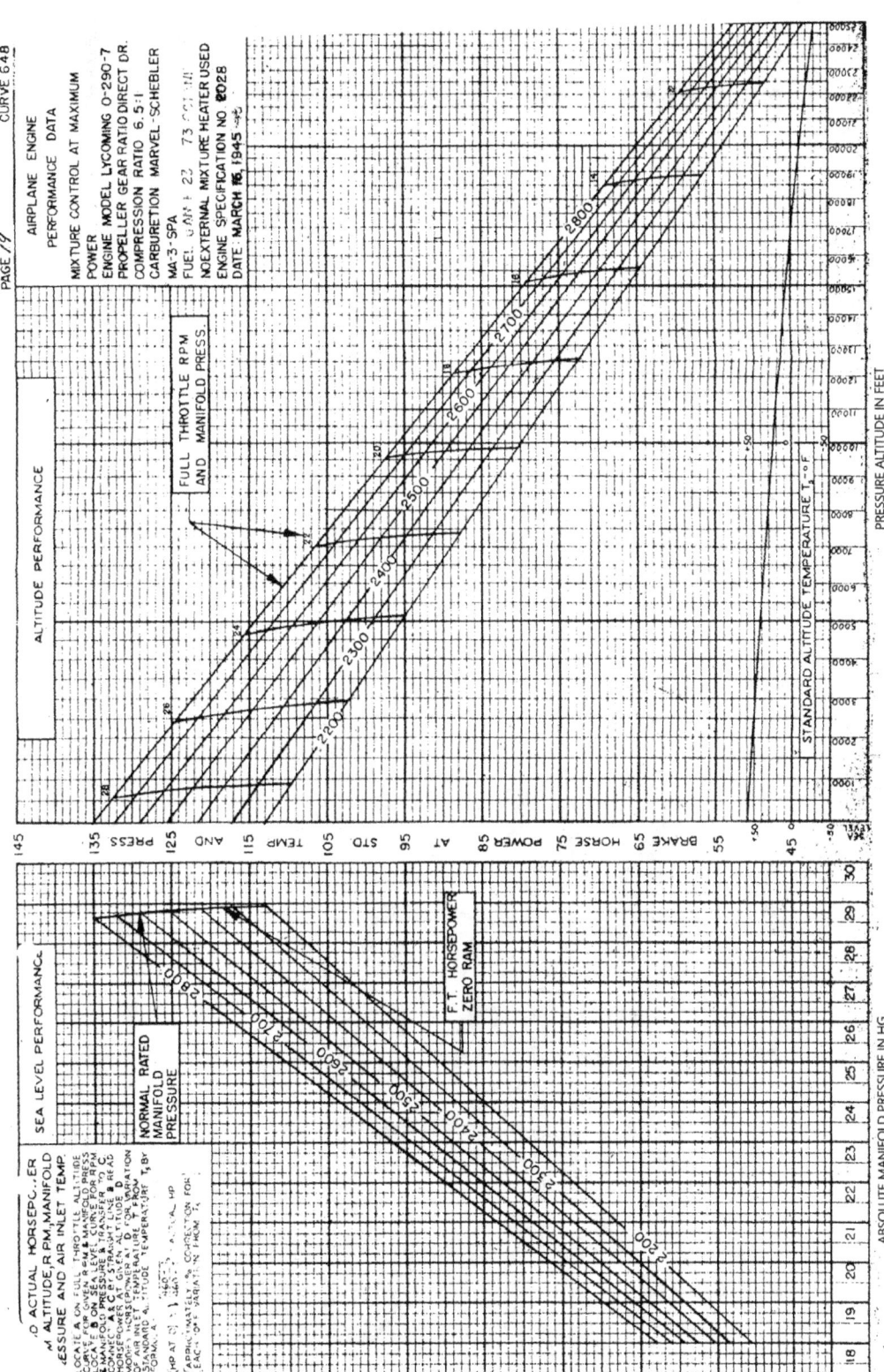

Lycoming O-290-7 Power Curve (NARA).

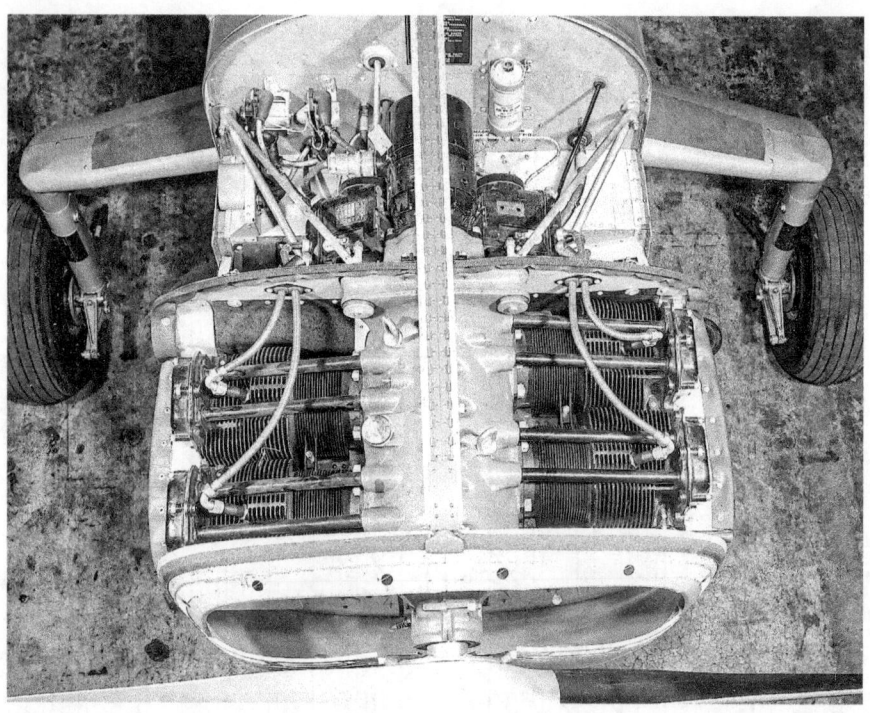

Top: Looking down on the O-290–7 installed on a YL-15 with the cowling and nose bug off (author's collection). *Bottom:* The Lycoming installed on the engine mount with the cowling removed (author's collection).

at will. A parallel set of linkages of adjustable control rods and bell cranks connected the pilot's throttle and mixture controls with the carburetor throttle valve and mixture lever.

The pilot's throttle control slot was marked "OPEN" at the front end and "CLOSED" at the aft end. The carburetor mixture control quadrant was marked "RICH" at the front and "IDLE CUT OFF" at the rear.

The XL-15 had good access for engine maintenance (author's collection).

XL-15 pilot's throttle and prop control (author's collection).

There was a large air inlet in the lower part of the cowl nose bug for carburetor air and cooling air. Engine baffles under the cowl and an adjustable cowl flap provided circulation around the engine in flight for cooling.

The engine cowl flap was actuated using a conventional push-pull flexible cable assembly with a knob control handle located on the lower right-hand corner of the circuit breaker panel in the cockpit. The cable ran through the firewall to an idler lever where a link led forward and down to the hinged cowl flap in the lower center cowl panel. The cowl flap had a hinge along its forward edge. Pulling the cowl flap knob opened the cowl flap, increasing the amount of air flowing through the engine compartment. The handle had a locking device to secure the cowl flap in the open or closed position.

The carburetor air intake system consisted of a welded aluminum alloy sheet metal housing containing an air filter, carburetor air box containing a valve for regulation of hot or filtered cold air, connecting to ducts to the exhaust heat exchanger, and necessary drain lines, screens, and fittings. An alternate cold non-filtered air inlet was provided in case the air filter became clogged.

The carburetor air heat control consisted of a flexible push-pull cable assembly extending from a control knob at the left-hand side of the instrument panel forward

8. Propulsion

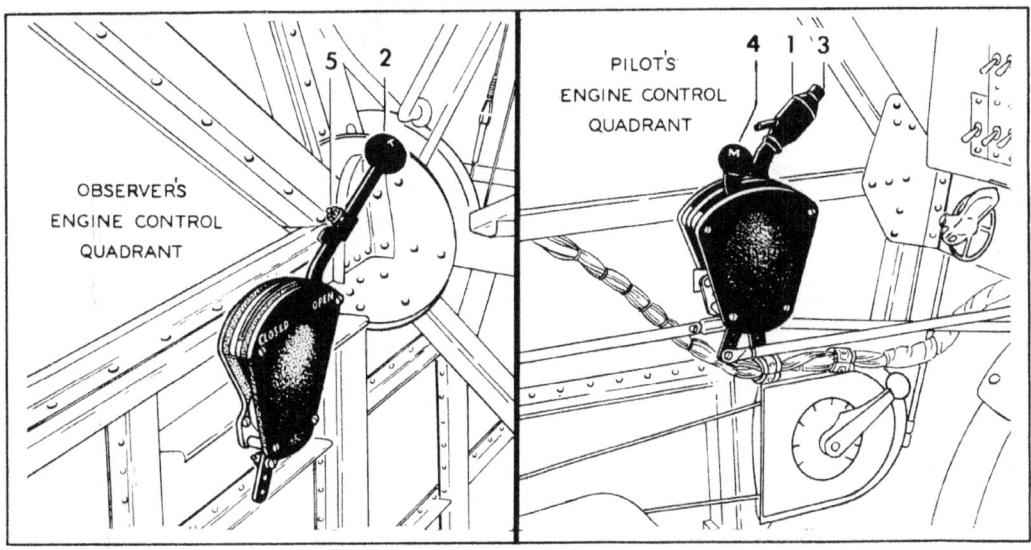

Engine Control Quadrants

1. Pilot's throttle control
2. Observer's throttle control
3. Microphone switch
4. Carburetor mixture control
5. Throttle control quick-release

Top: XL-15 observer's throttle control (author's collection). *Bottom:* YL-15 Pilot and observer's engine controls (author's collection).

through the firewall to a valve on the left-hand heat exchanger, which was installed around the engine exhaust manifold. The carburetor air heat control could be locked in position by rotating the knob clockwise. When the control knob was pulled out, heated air was directed into the carburetor intake to prevent icing, or when pushed in, the carburetor intake received ram air. The carburetor air heat should only be used full-on or full-off and never in a partial-on position.

The engine was of the pressurized wet sump type, and there was no external oil tank or oil cooler. The filler neck and dipstick were located on the upper-right-hand corner of the engine case and were accessible through a hinged door in the right-hand hinged cowl panel. All engine parts were constantly bathed in oil circulated by the engine oil pump. There were oil pressure and temperature gauge connections. Specification AN-O-08 engine oil was to be used with Grade 1100 for ambient temperatures above 40° F and Grade 1080 for lower temperatures.

The engine was equipped with two Bendix-Scintillia SF4LN-8 magnetos located at the upper rear of the engine in symmetrical right- and left-hand positions. Access was gained through the hinged engine cowl panels. Each magneto operated on the principle of a rotating magnet operating inside of stationary coil windings. The magnetos provided a high-tension spark to each cylinder through two Auto-Lite SH-2K

It was easy to check the engine oil (author's collection).

aircraft spark plugs per cylinder. Both sparks occurred simultaneously in each cylinder at 20° before top dead center. The firing order was 1–3–2–4. The magneto windings were enclosed in a housing to protect them from moisture and vibration. The rotating magnet was of two-pole construction and was made of high-tension magnetic steel, which enabled it to maintain a stable magnetic field for long periods. The rotating magnet operated on two bore bearings, one located at the breaker end and the other at the drive end. A conventional set of contacts was housed at the rear of the magneto and was operated by a tooled cam. Arcing at the contacts was held to a minimum by using a cartridge-type condenser which could be removed and replaced whenever necessary. The magnet shaft rotates at engine speed and had a small distributor gear mounted on it which drove a large distributor gear at one-half engine speed. A cylindrical rotor was mounted on the latter gear and functioned to distribute the high-tension current to the correct engine cylinder at the proper time. Both magnetos were driven through an impulse coupling that provided a hot spark for starting and retarding the ignition timing to eliminate kickbacks when starting. When the engine was running, the impulse couplings were inoperative. Both the magnetos were cooled by air circulated from the forward compartment of the engine cowling by a small tube that directed the air into a vent on the side of each magneto. On the lower left-hand side of the pilot's control panel was a four-position

A view of the accessories mounted on the rear of the engine (author's collection).

YL-15 left-hand exhaust stack (author's collection).

lever handle ignition (magneto) switch and dial marked "BOTH," "L," "R," and "OFF."

The engine had an Eclipse type 756-56-C starter with a series-wound electric drive motor having reduction gearing, overload torque release, automatic engaging device, and drive jaw. The overload torque release consisted of a multiple disc clutch with provision for adjusting the clutch to any required torque fit. The starting motor functioned only to crank the engine for starting. It was mounted at the rear of the engine in alignment with the engine crankshaft, and access to it was gained by raising the hinged engine upper cowl panels. Located just to the right of the magneto switch on the control panel was a momentary-position spring-loaded toggle starter switch that was protected by a red guard that must be lifted before the switch could be closed. The starter switch was open when the toggle was upright and closed when pressed against the spring.

The generator used on the YL-15 was a standard-type M-3 built to AAF Specification No. 95-3236.1. It had a rating of 50 amperes at 25.5 volts and 2500 to 4500 rpm. The generator had a four-bolt mounting flange and weighed 20 pounds. It was designed to supply electrical power for operating the electrical equipment in the airplane and to keep the battery fully charged. A conventional carbon-pile voltage

YL-15 right-hand exhaust stack (author's collection).

regulator regulated it. The generator was mounted at the rear of the engine, and access to it was gained by raising the hinged engine cowl panels.

The exhaust system consisted of left and right welded stainless steel type collectors supported from the engine cylinders. A muff-type heat exchanger was installed around each exhaust collector to provide heated air to the carburetor and the cabin.[2]

Propeller

The YL-15 was equipped with a McCauley two-bladed fixed-pitch metal propeller model 1A170-L8043. The propeller diameter was 80 inches with a blade angle of 13.25° at a 75 percent radius. The propeller had a weight of 33.25 pounds. The propeller was directly connected to the engine drive shaft with eight lock wired studs and nuts.

Ground clearance was 15.4 inches on a level landing. The distance between the rear face of the propeller and the engine cowling was 4 inches.

Many different propellers were tried on the two XL-15 airplanes. The *Pilot's Handbook* for XL-15 sn 46–521 lists the propeller as a Sensenich Brothers constant-speed propeller. The pitch control was on a modified type B-17 engine

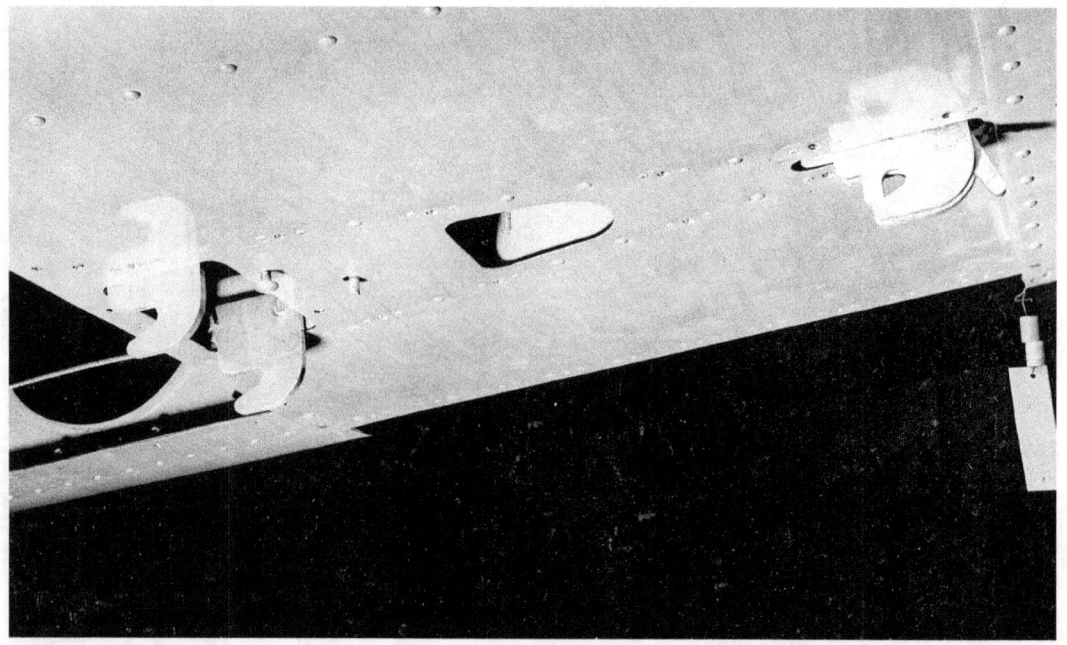
The JATO bottle was mounted to the drop tank attachment points (author's collection).

control unit. The control actuated the propeller hydro-control lever mounted on the rear of the engine. The control was set in the "LOW" position for takeoff and landing. After takeoff, a cruise condition could be set up and the propeller pitch control set in the "HIGH" position.[3]

JATO

The Boeing L-15 bid specification states that Jet Assisted Take Off may be used for takeoff. The takeoff over 50-foot performance estimates for the Boeing bid showed 325 feet when JATO was used compared to 595 feet normally, a reduction of 270 feet. This was estimated using an Aerojet 8AS-1000E JATO unit that produced 1,000 pounds thrust for 8 seconds and weighed 140 pounds. The JATO bottle was mounted to the same attachment points used for the drop tank on the lower side of the fuselage.[4]

Air Tow

The L-15 was designed so that it could be towed from another aircraft. When being air towed the propeller was locked. Early design sketches of the aircraft under tow showed that the tow hook would be below the fuselage cabin. It appeared in these that possibly the forward JATO/drop tank attachment hook was to be used for attaching the towline. It was stated that the tow line could be released, and then the engine started in the air to resume normal flight and landing.[5]

The XL-15 *Pilot's Handbook* talked of a different arrangement. It stated that as part of alternate equipment, a kit would be provided consisting of a bracket for attachment of an aerial tow release mechanism and propeller locking mechanism together with the necessary operating system. This equipment would not be installed in the airplane as delivered. Towing in flight was to be accomplished by locking the propeller in the horizontal position and attaching the tow line to towing release fittings at each outboard end of the landing gear sponson tubes. The propeller lock control handle was located on the lower left of the instrument panel and provided a safety shield over the starter and ignition switches to prevent engaging the starter with the propeller locked.[6]

Flight testing of the aerial tow kit was done on YL-15 47–423 in April 1950.[7]

9

Landing Gear

The Boeing L-15 was designed with landing gear provisions to operate from land, snow, or water. It could also be equipped for operating from a Brodie system on land or a ship.

Land

The L-15 had a tail wheel landing gear. However, the main gear portion was anything but conventional. The main landing gear consisted of two main landing wheels supported by a cantilever spanwise sponson tube attached to the underside

The L-15 had a tailwheel landing gear mounted at the back of the gondola (author's collection).

9. Landing Gear

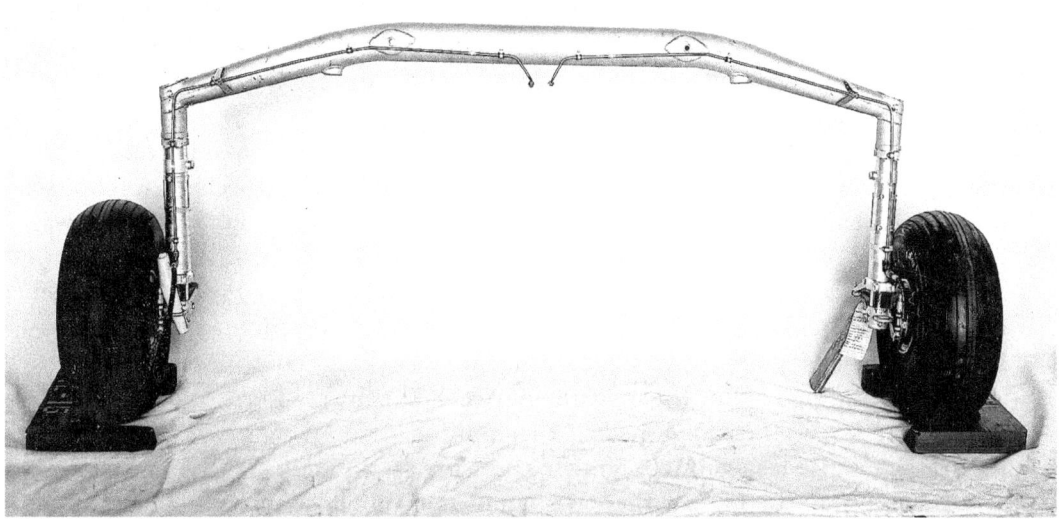

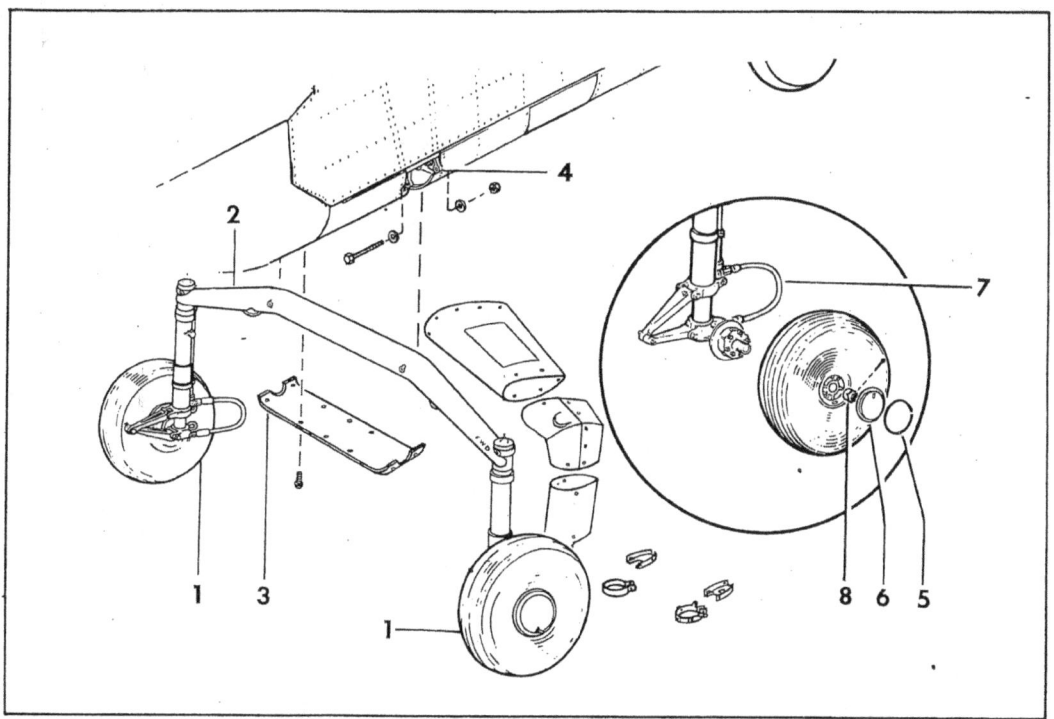

Removal of Landing Wheels

1. Landing wheel
2. Sponson tube
3. Sponson gap closure
4. Hub
5. Locking ring
6. Cover plate
7. Brake hose
8. Axle nut

Top: The main landing gear consisted of two wheels supported by a cantilever spanwise sponson tube (author's collection). *Bottom:* Components of the main landing gear (author's collection).

of the fuselage. The main wheels of the YL-15 were supported by spring-hydraulic shock absorbers installed in tube cans welded to the outboard ends of the sponson tube. The XL-15 used air-oil shock absorbers.[1] Streamlined sheet metal fairings covered the sponson tube and shock absorbers. Bolts inserted through holes in the sponson tube attached it to fittings on the lower fuselage at FS 75.

Each main landing wheel was equipped with a type 111, non-skid tread, 4-ply rayon cord 8.50 × 6.0 in. ten psi low-pressure tire and tube. The wheel hub was bolted to the lower end of the shock absorber piston. The stroke of the shock absorber was 7.03 inches, allowing for vertical wheel travel of 7.0 inches.

The wheels had hydraulic landing wheel (expander type) brakes with master cylinders, compensating cylinders, and reservoir, which were actuated by toe pressure on the pilot's rudder pedals. A parking brake (locking) control was actuated by a tee handle located below the instrument panel. The brakes were locked by applying toe pressure on the rudder pedals, then pulling out on the tee handle. The parking brake was disengaged by applying toe pressure on the pedals.

On each sponson fairing, there was a non-skid rubber step pad for entering the cockpit and servicing the fuel tanks. There was an access panel on each sponson tube fairing to fill the oil in the shock absorbers.

The main landing wheels could be rotated to shorten the wheel tread from 87.7 inches to 61.7 inches in order to load into a standard Army deuce and half truck. This was accomplished by removing the upper scissor arm attachment bolt from the lug on the lower end of each shock absorber cylinder. This lug was on the forward side of the cylinder, and there was another lug on the aft side. The wheel could be swiveled around to the inside to secure the scissor arm to the aft lug.

The tail wheel was not at the tail but instead was attached to the rear of the fuselage gondola by the tail wheel drag strut, which held a tail post assembly positioned on the airplane centerline at FS 162. This post, in turn, held both the spring-hydraulic shock absorber that supported the tail wheel and the steering mast by which,

By removing a scissor arm attachment bolt, the wheels could be rotated to allow loading into a standard Army two and a half ton truck (author's collection).

through cables, the tail wheel could be steered by operation of the rudder pedals. There were turnbuckles on the cables for adjustment. A cam in the tail wheel post disengaged the control system when the tail wheel went beyond the arc of rudder travel plus 5° in either direction; thus, it became free swiveling. The tail wheel was equipped with a type 111, non-skid tread, 5. 00 × 4. 0 in. 30 psi low-pressure tire and tube. The tail wheel had a total travel of 8.5 inches.

A filler cap was located on top of the tail wheel shock absorber and could be accessed through a hinged access panel located on the centerline of the observer's compartment floor near the doors.[2]

Both the main landing gear and tail wheel were drop tested.

Snow

The L-15 could be equipped with skis for operation off of snow. The skis were attached to the standard landing gear axles in place of the wheels and were quickly and easily interchangeable with the wheels. Three skis were used, one each in place of the main wheels and a smaller ski in place of the tail wheel. The skis were of metal construction.[3]

Skis were evaluated on the second XL-15 prototype (46–521).

Water

The L-15 could be equipped with twin floats for operation off of water. The floats were Edo Model 44–2425 floats, which were 17.00 feet long and had a 29-inch beam. They were mounted on a truss-type structure constructed with 2.25-inch streamline tubing that replaced and tied into the attachments on the fuselage for the standard

XL-15 on skis (author's collection).

Top: The skis fit on the axles in place of the wheels (author's collection). *Bottom:* XL-15 being tested on floats. Note the flaps are down (author's collection).

XL-15 on Edo floats (author's collection).

landing gear. The floats had water rudders connected through cables and pulleys to the rudder petals. The float spread centerline to centerline was 94 inches.[4]

A 0.08 scale model of the XL-15 with Edo 44–2425 floats was tested in the University of Wichita's four-foot-diameter wind tunnel. The wind tunnel test was used to support structural, performance, stability, and control analysis of the floatplane. The wind tunnel tests indicated that the elevator power was not adequate for takeoff with flaps zero. At 20° flap setting, there was sufficient elevator power, and the takeoff distance on the water was estimated to be 760 feet with a takeoff time of 19.6 seconds.[5]

The gross weight of the XL-15 with floats was 2,260 pounds. Each float weighed 120 pounds and the water rudder weighed 8 pounds. The spreaders between the floats weighed 39 pounds. and the attachment struts and wires weighed 52 pounds.[6] Thus, the useful load was substantially reduced with floats.

The second XL-15 prototype (46–521) was fitted with floats.

Brodie System

The Brodie system was a device that permitted takeoff and landing from a 300- to 500-foot length of cable about 65 feet above the surface, which enabled light aircraft operation from a ship or on land where there was no runway. During World War II, Army and marine pilots operated L-4s and L-5s off of LSTs equipped with

Top: Maintenance being performed while standing on a float (author's collection). *Bottom:* Brodie system mounted on an XL-15 (author's collection).

Brodie systems to provide artillery fire direction during amphibious operations. The Brodie system was also used in the China-Burma area in rough terrain where an airstrip could not be built.

The aircraft was fitted with a hook mechanism mounted on top of the cockpit sticking up about 6 feet. To understand how the Brodie system was used, I refer to Don Baker, who wrote on May 11, 1998, in rec.aviation.military, "The Brodie system was necessary because our L-4's only had limited range, and if they were launched from aircraft carriers or other ships well offshore we would use up so much of our fuel just getting to the beach that we would not have been able to remain there for very long to adjust naval, artillery, or Air Force fire before we would have had to return for fuel. So we were launched and recovered right near shore from specially modified LST's. A cable and pulley system was strung down the side of the LST, and the L-4 Cub, with a hook on top, was hoisted up onto the cable and hung from a loop, dangling in mid-air. The engine was started, revved up, and the airplane was released, running down the cable to gain flying speed. The pilot then disengaged it from the cable and flew off on the mission. The recovery was more tricky: we had to fly the hook on the top of the airplane back into the loop on the cable, and if successful, were braked to a stop on the cable, dangling in the air again, and then they lowered the airplane onto the deck of the LST."[7]

Brodie gear was mounted to the top of one of the prototype XL-15s, but it is not known if flight tests of the Brodie system were actually performed.

10

Systems

Fuel System

The YL-15 had a fuel system consisting of wing fuel tanks, fuel strainer, fuel cocks, controls, primer, and the necessary lines and fittings. Fuel was fed to the engine by a gravity feed system from the wing tanks through a selector valve and strainer to the engine carburetor. There was also a provision for an external drop tank. AN-F-48 73 octane gasoline was the prescribed fuel in most documents; however, the March 23, 1949, Handbook Flight Operating instructions recommended AN-F-48 80 octane gasoline with an alternate of U.S. Army Specification No. 2–103, grade 80 octane general purpose (truck) fuel.

A 10.5-gallon rubberized nylon crash-resistant bladder cell fuel tank was located in each wing panel. The fuel filler caps were under the filler access doors on the upper surface of each wing leading edge, close to the fuselage. Installed on the upper surface of each tank were the filler cap and scupper, vapor vent line, and

Fuel filler in the upper wing surface (author's collection).

Liquidometer transmitter. Each fuel tank had a sump made from cast aluminum alloy with bosses for a drain cock, AN4007–2 finger strainer, and fuel line attachment. The fuel strainer could be removed without disturbing the fuel line. Access to the sump and drain was by a circular door on the lower surface of each wing. A fuel shut-off cock located in the wing to fuselage gap was provided in the lines to permit removal of the wing without draining the tank.[1]

At the lowest point in the line from the fuel shut-off valve to the carburetor, a fuel strainer was accessible from outside the aircraft without removing the cowl, and it drained clear of the aircraft. There was a line running aft from the strainer to the primer valve. The manual primer control consisted of a primer valve lever handle marked "ON" and "OFF" and a push-pull tee handle to the primer pump located on the lower right-hand side of the pilot's instrument panel. Lines ran forward from the primer valve to nozzles in the fuel intake port of each engine cylinder. The lever was rotated up to the "ON" position, and then the engine was primed with fuel by pumping the tee handle out and in. The tee handle was pushed full in when the engine started, and the lever rotated down to the "OFF" position.[2]

The fuel selector valve was attached to the aft face of the firewall on the left-hand side, and a shaft attached it to the four-position lever handle and indicator dial installed on the lower left-hand side of the pilot's instrument panel. The four positions of the valve were "LEFT WING 10. 5," "RIGHT WING 10. 5," "OFF," and "AUX." The auxiliary fuel tank pump switch had to be turned on before turning the fuel selector valve to "AUX." Note that if the auxiliary tank was not installed, there was a plate labeled "OFF" to install over the lower right corner of the fuel selector dial to cover the "AUX" label.[3]

On the right-hand side of the instrument panel were an AN3106–14S-5S dual fuel gauge and an AN3157–6 fuel warning light. The fuel gauge had two scales and pointers, marked "L" and "R," which indicated the gallons of fuel in the wing tanks. The gauge received an electrical signal from Liquidometer units installed in each wing tank. The Liquidometer unit translated a float position to an electrical signal. The red fuel warning light was wired to the fuel gauge and came on when approximately 20 minutes of fuel remained in the right-wing tank. Because of the wiring of the low fuel warning light, it was recommended the fuel in the right-wing tank be used last.[4]

The auxiliary tank carried 25 gallons (earlier documents said 24 gallons) and was of a streamlined shape made of aluminum. It was attached to three mounting hooks on the underside of the fuselage (also used for mounting a JATO bottle) and was jettisonable by pulling a lever on the lower left-hand side of the pilot's cockpit floorboard. An electric fuel pump was mounted below the cockpit floorboard to pump the fuel from the auxiliary tank through a fuel line to the fuel selector valve. The auxiliary fuel pump had to be switched "OFF" before pulling the tank jettison handle.[5]

If the auxiliary fuel tank was installed, the standard operating procedure was to use from the auxiliary fuel tank first, the left tank next, and the right tank last. When running from the auxiliary tank with fuel in the wing tanks, if the engine

Auxiliary fuel tank (author's collection).

began to sputter and rpm dropped, the procedure was to switch immediately to one of the wing tanks, preferably the left, and to switch the auxiliary fuel pump off.[6]

Flight Control System

The flight controls of the YL-15 were divisible into six separate systems: elevator, elevator tab, rudder, tail wheel, spoiler aileron, and flap. These are described in subsequent sections. All the movable surfaces were controlled from the cockpit by the control stick, rudder pedals, flap control lever, or elevator tab control handle. A second detachable control stick and additional rudder pedals in the observer's compartment permitted the observer to take over sufficient control for emergency operation of the plane. Any time the rear seat was not in use, the observer's control stick should be in the stowed position. Each of the six systems was operated through control cables, bell cranks, pulleys, and control rods, and each incorporated disconnect

Attachments for auxiliary fuel tank on the lower fuselage (author's collection).

fittings, turnbuckles, and fairleads, where necessary. Access doors were provided for removing components and adjusting the systems. Repairs to surface controls were not normally undertaken since these parts were to be replaced if damaged, but maintenance inspection, at frequent intervals, for frayed cables, worn pulleys, chafed fairleads, and the like was required. Cables had to be kept well coated with Paralketone (a corrosion inhibitor). Bonding braids had to be replaced, and turnbuckles had to be always safety wired whenever the original wire had been cut. Cable tension was to be accurately maintained and checked by the use of a standard tensiometer. Defined control surface angular travel was subject to a 2° tolerance in either direction.[7]

Elevator Control System

The elevator control system was a linkage between the dual control sticks and the control horn on the elevator. The purpose was to convert forward or aft movement of the control sticks in the cockpit into up or down rotation of the elevator. There were two elevator stop bolts in the pilot's control stick mechanism, which were adjustable to set the proper limits of stick travel. The elevator control linkage consisted of the elevator up cable, the elevator down cable, an elevator bell crank located in the rear end of the tail boom, and an adjustable control rod installed in the horizontal stabilizer. The forward ends of the up and down cables attached to the pilot's control stick, with the elevator up cable running to a pulley aft of the pilot's seat while the elevator down cable ran forward to a pulley in front of the stick, then around it and back to a pully aft of the pilot's seat. Both cables then went up to pulleys at the top of the cabin then ran aft through the tail boom to the elevator bell

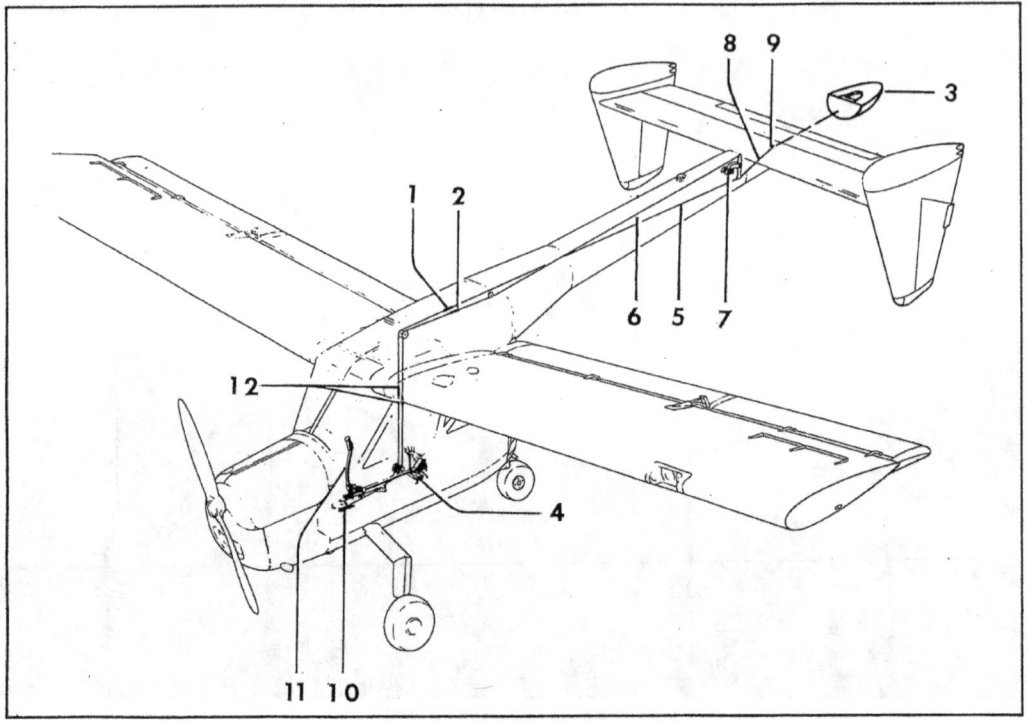

Elevator Control System

1. Loop disconnect
2. Spring disconnect
3. Tail boom fairing
4. Elevator bellcrank
5. Elevator down cable
6. Elevator up cable
7. Elevator balance weight
8. Elevator control rod
9. Elevator horn
10. Idler pulley
11. Control stick
12. Turnbuckles

Top: Elevator control system (author's collection). *Bottom:* Elevator control rod transmitted motion from the bell crank to the elevator horn (author's collection).

crank. An elevator balance weight was attached to the bell crank to offset the down drag due to the elevator weight. A control rod transmitted movement from the bell crank to the elevator horn. There was a loop disconnect in the down cable and a spring disconnect in the up cable, which were accessible from the observer's compartment and allowed for disassembly of the tail boom from the fuselage.[8]

A torque tube ran aft from the pilot's control stick to a socket for the observer's movable control stick. A quick release on the observer's control stick permitted easy removal for stowage when not in use on a bracket on the ceiling of the observer's compartment.[9]

Elevator Trim Tab Control System

The elevator trim tab control system transmitted the manual tab control movement to the elevator tab and held the tab in the position established to trim the aircraft.

The elevator tab control system consisted of a manual control at the left-hand side of the cockpit directly below the engine control quadrant. The tab control was a short lever with a knob grip attached to a pulley in an enclosed bracket. The face

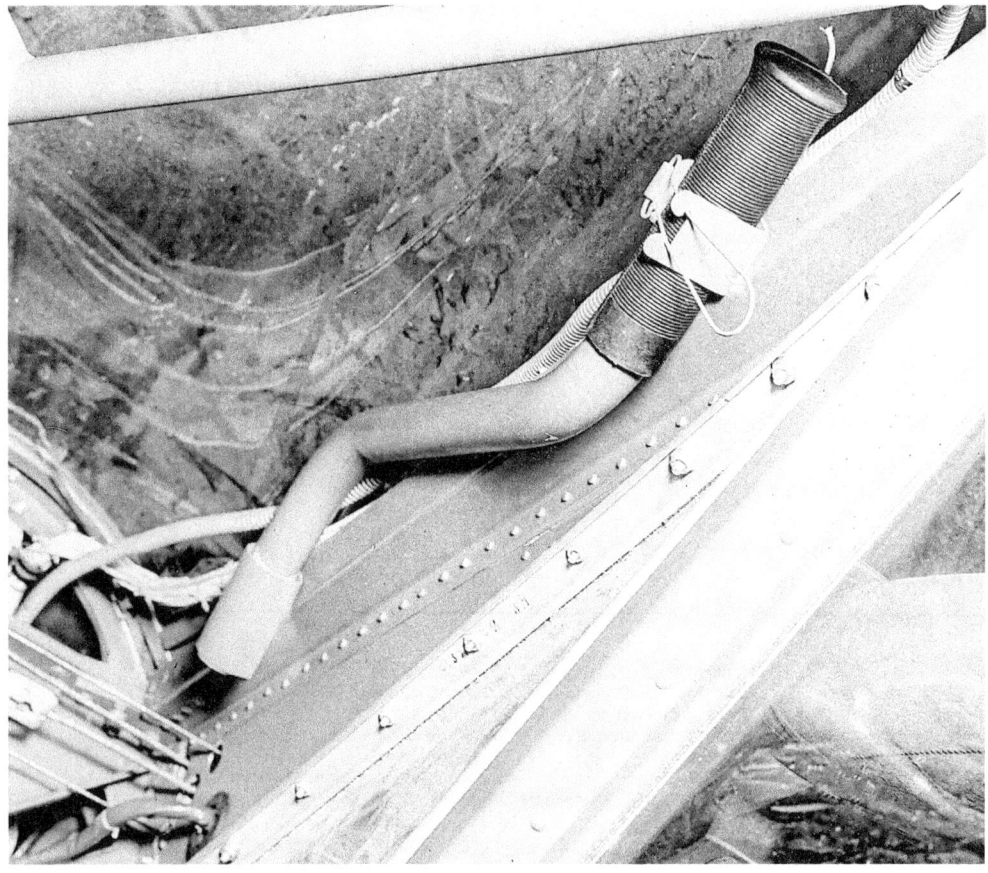

Stowed observer's control stick (author's collection).

of the bracket was marked with positions from 0° to 20° and with the indications "NOSE UP" and "NOSE DOWN." The lever was straight up at the "0" position. Pulling the lever aft caused the tab to defect down or "NOSE UP" and pushing the lever forward deflected the tab up or "NOSE DOWN." Cables led aft from this control to a gearbox in the horizontal stabilizer, and adjustable control rods connected the gearbox with the elevator tab. Two cables, an elevator tab up cable and an elevator tab down cable, passed over pulleys and through fairleads aft from the control to a point behind the pilot's seat and then upward to the top of the fuselage, then aft through the observer's compartment and tail boom to the gearbox. A handle with a ball grip operated the control. Moving the handle forward or aft took up on one cable and paid out on the other cable. This cable travel rotated a worm gear in the gearbox, which actuated a worm gear attached to an arm extending out from the gearbox. Two control rods through the elevator converted forward and aft movement of the arm to up and down deflection of the elevator tab.[10]

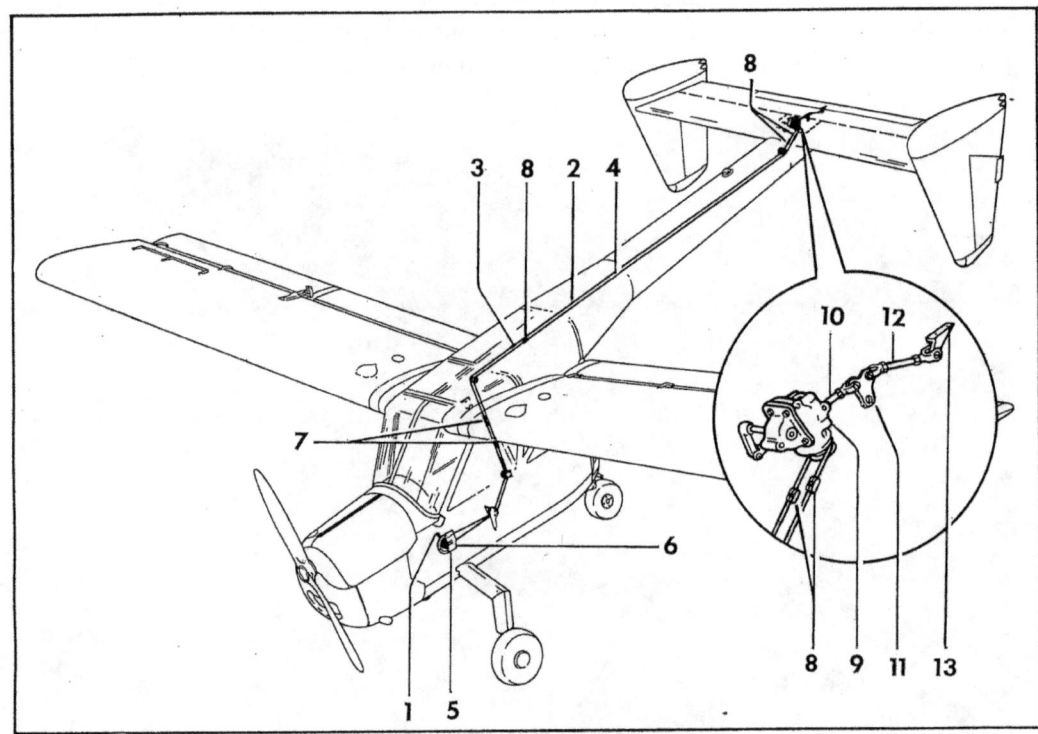

Elevator Tab Control System

1. Control handle
2. Elevator tab down cable
3. Rigging disconnect hook
4. Elevator tab up cable
5. Bracket
6. Control drum
7. Turnbuckle
8. Loop disconnect
9. Gear box
10. Forward control rod
11. Idler lever
12. Aft elevator tab control rod
13. Elevator tab horn

Elevator tab control system (author's collection).

Rudder Control System

The rudder control system transmitted the movement of the rudder pedals in the cockpit or observer's compartment into left and right movement of the dual rudders.

The rudder control system consisted of a dual set of cables connecting the pilot's and observer's rudder pedals with the left and right rudders. A rudder pedal quadrant was installed on each of the observer's two pedals, and they were connected by left- and right-hand torque tube assemblies to the pilot's pedals so that they moved in unison. A balance cable led forward around two pulleys connected the observer's rudder pedals so that the left- and right-hand pedals moved simultaneously but in opposite directions. From the quadrant on each observer's pedal, a single cable led upward to a turnbuckle, where it was separated into a left rudder cable and a right rudder cable. These cable pairs passed over pulleys through the observer's compartment and through the tail boom into the horizontal stabilizer. There the left rudder cable of each pair led to the arm on the left-hand rudder bell crank, and the right rudder cable to the right-hand rudder control rods transmitted bell

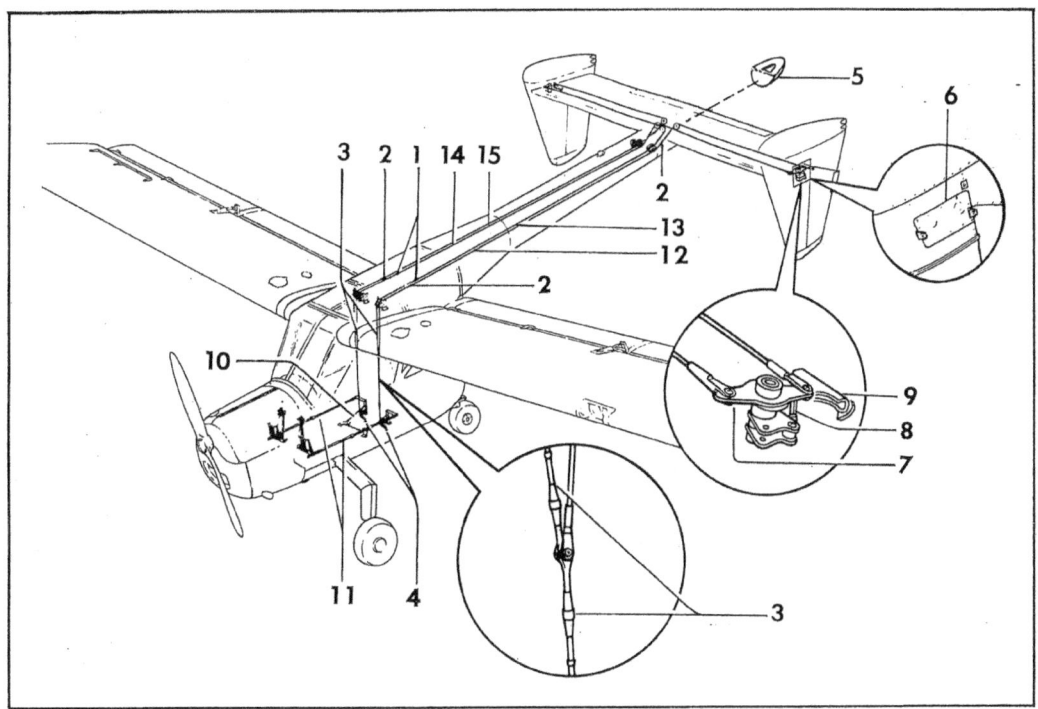

Rudder Control System

1. Spring disconnect
2. Loop disconnect
3. Turnbuckles
4. Rudder pedal quadrants
5. Tail boom fairing
6. Fin access door
7. Rudder bellcrank
8. Rudder control rod
9. Rudder horn
10. Balance cable
11. Rudder pedal torque tube
12. Left-side left rudder cable
13. Left-side right rudder cable
14. Right-side left rudder cable
15. Right-side right rudder cable

Rudder control system (author's collection).

crank movements aft to the respective rudder horns. Loop and spring disconnects were installed in the cables where they passed through the observer's compartment, and loop disconnects, where the cables passed from the tail boom to the horizontal stabilizer. Standard pulleys, pulley brackets, and fairleads were provided where necessary.[11]

Adjustment lugs (stops) on the inboard lower corners of the pilot's rudder pedals allowed adjustment of the pedal position for the pilot's leg length. They could be adjusted one at a time or both together. The adjustment was by pressing inboard against the stop and simultaneously pushing the pedal into the desired position and then relieving foot pressure from the stop to lock the pedal in the position attained. The observer's rudder pedals were not adjustable.[12]

There was a ground adjustable trim tab on the left rudder.

Tail Wheel Control System

The tail wheel control system was designed to permit the airplane's ground steering using the rudder pedals, rather than using wheel brakes in a conventional manner. However sharper turns than were possible with this restricted pedal

Rudder horn and bell crank (author's collection).

movement could be accomplished using the landing wheel brakes, but if this was done, the tail wheel automatically disengaged from its control system and became swivelable.[13]

The tail wheel control system consisted of two cables having turnbuckles for length adjustment. These connected the rudder pedals with a steering mast on the tail wheel post. Rudder pedal movement thus deflected the tail wheel in conjunction with the rudders. A cam in the tail wheel post disengaged the control system when the tail wheel deflected, in either direction, an amount equal to the rudder travel arc plus 5°. The tail wheel was then swivelable through the balance of the complete arc.[14]

Spoiler Aileron Control System

The spoiler aileron control system raised or retracted the spoiler ailerons as the control stick was moved sideways. When the control stick was

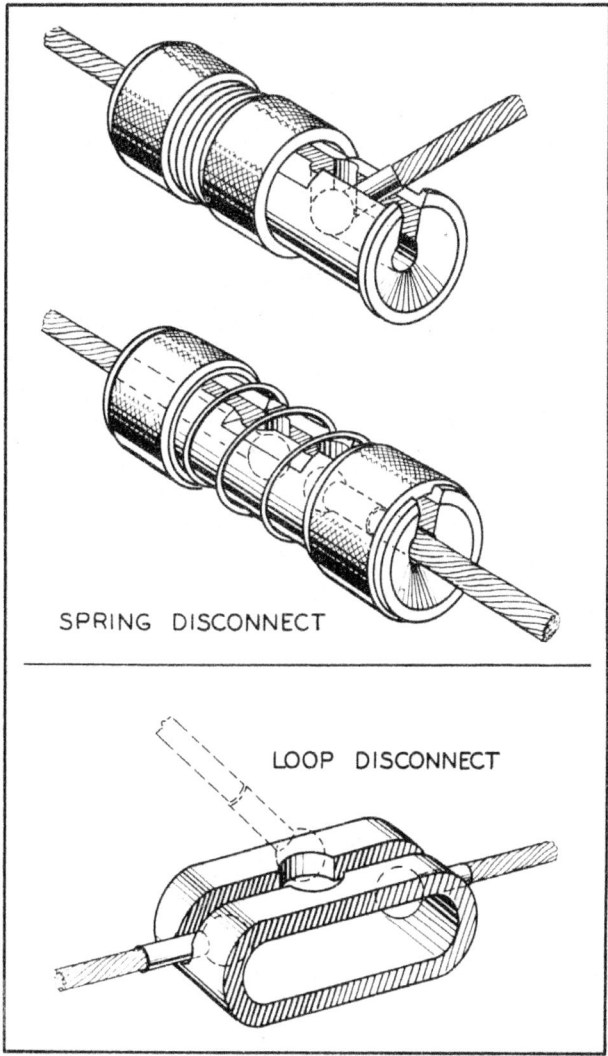

Control cable disconnect fittings (author's collection).

moved to the left, this system caused the left-hand spoiler aileron to raise up; when the control stick was centered (neutral), both ailerons retracted; and, when the control stick was moved to the right, the right-hand aileron raised up. The spoilers augmented the aileron effect accomplished by the differential flap movement.[15]

The spoiler aileron control system incorporated a single cable that led from a control horn on the aileron and elevator central control mechanism, up and outboard around quadrants in the two wing panels. A lever-type quick disconnect, known as the rigging disconnect hook, was used to join the two ends of the cable. There were adjustable control rods that connected the quadrants to the spoiler ailerons. The horn was linked to the control stick so that when the control stick moved laterally, the spoiler ailerons rose above, or retracted below, the wing upper surface. Standard pulleys, pulley brackets, and fairleads were provided where necessary.[16]

Flap Control System

The flap control system served a dual purpose. It converted forward and aft movement of the flap control lever into up and down movement of both flaps, moving in unison for lift control. It also transmitted lateral movement of the control stick to the flaps, so differential flap position was varied to give an aileron effect.

These two movements operated independently of each other. When the control stick was laterally centered in the neutral position, the two flaps moved up or down together in the same relative position. If the control stick was moved to the left, the right flap lowered, and

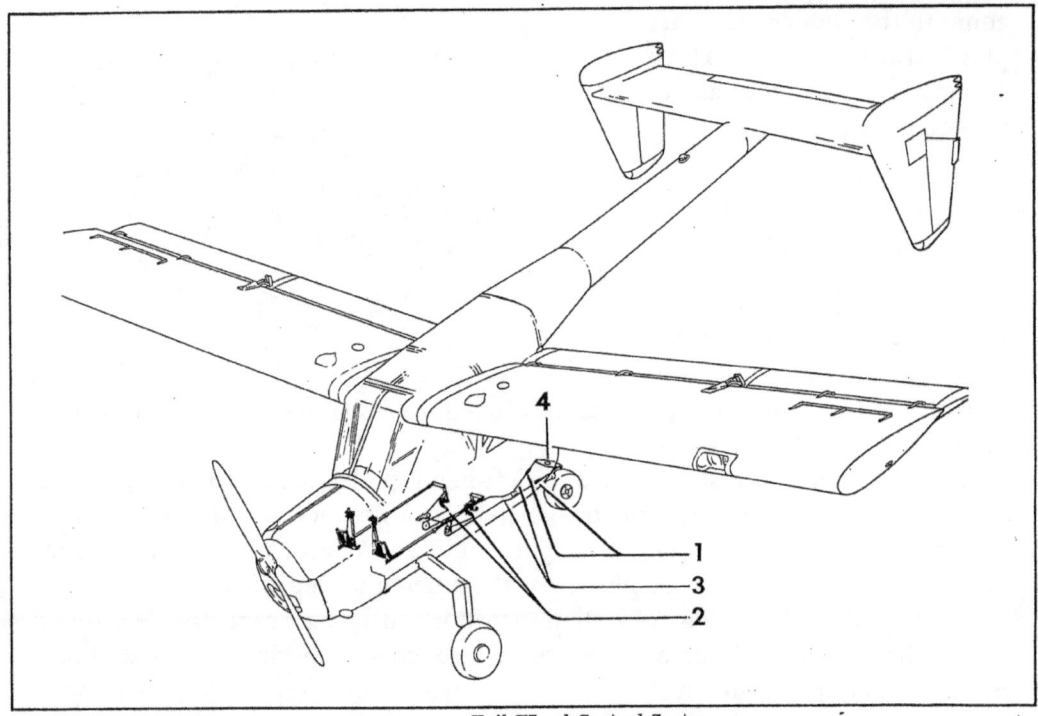

Tail Wheel Control System
1. Turnbuckle
2. Rudder pedal quadrant
3. Tail wheel cable
4. Tail wheel steering mast

Top: Bent metal rudder tab (author's collection). *Bottom:* Tailwheel control system (author's collection).

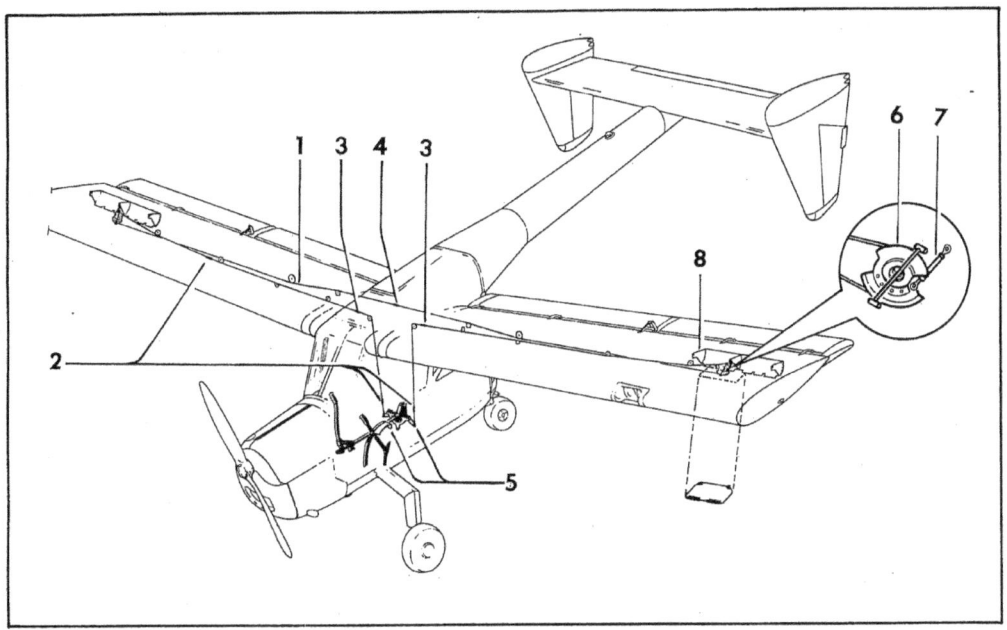

Spoiler Aileron Control System
1. Spoiler aileron cable
2. Turnbuckle
3. Loop disconnect
4. Rigging disconnect hook
5. Spoiler aileron control horn
6. Spoiler aileron quadrant
7. Spoiler aileron control rod
8. Spoiler aileron

Spoiler aileron control system (author's collection).

the left flap raised to give a greater drag and consequent rolling moment acting as an aileron.[17]

The flaps were moved by the flap control lever mounted in a quadrant at the left-hand side of the pilot's seat. The flap control lever's forward and aft movement was transmitted by the system and converted into the up and down movement of the flaps. There was a thumb-operated catch on the lever handle that stopped handle motion at selected flap positions. There was a positive stop at the neutral (Flap 0) position. When the lever was at this forward stop, a downward push, together with a twice repeated pressure on the thumb catch, allowed the lever a slight overtravel which gave a 10° up flap position for purposes of maneuvering and taxiing.[18]

The flap lever through cables and pulleys moved a bell crank, located in the center fuselage below the observer's control stick, linked by four cables to flap bell cranks in each wing. Flap control rods connected the flap bell cranks to control arms on the flaps. Pulleys, turnbuckles, cable disconnects, and fairleads were installed in the system where necessary.[19]

The center bell crank was also connected to the elevator torque tube on the aileron and elevator central control, and thus, lateral control stick movement affected the relative position of the flaps, allowing them to function as ailerons. In addition, the spoiler aileron control cables were connected to this bell crank. This connection provided for the operation of the spoiler ailerons in conjunction with the flap aileron action.[20]

Pitot-Static System

The pitot-static system of the YL-15 consisted of two static ports, a pitot tube, the necessary tubing lines, and pressure-actuated instruments. The pressure-actuated instruments were the airspeed indicator, rate of climb indicator, and altimeter.

The pitot tube was L-shaped and mounted underneath the left wing outboard of the fuel tank at approximately wing station 60 and behind the rear spar. It was an AN5812–1 pitot tube that incorporated an electrical heater with a control switch on the light switch panel on the instrument panel. The pitot tube sensed the total pressure, which passed through a tubing line from the pitot tube inboard to the fuselage frame, then forward and down to the area behind the hinged instrument panel, then to the total pressure port on the back of the airspeed indicator. The pitot tube connections to the tubing and electrical heater wire were accessible through the trailing edge access door adjacent to it. The tubing line had a quick-release fitting installed in the left-hand wing to body gap to allow breaking the pitot tube line when the wing panel was to be removed. There was also a drain cock at the low point in the tube to clear the line of condensed moisture.[21]

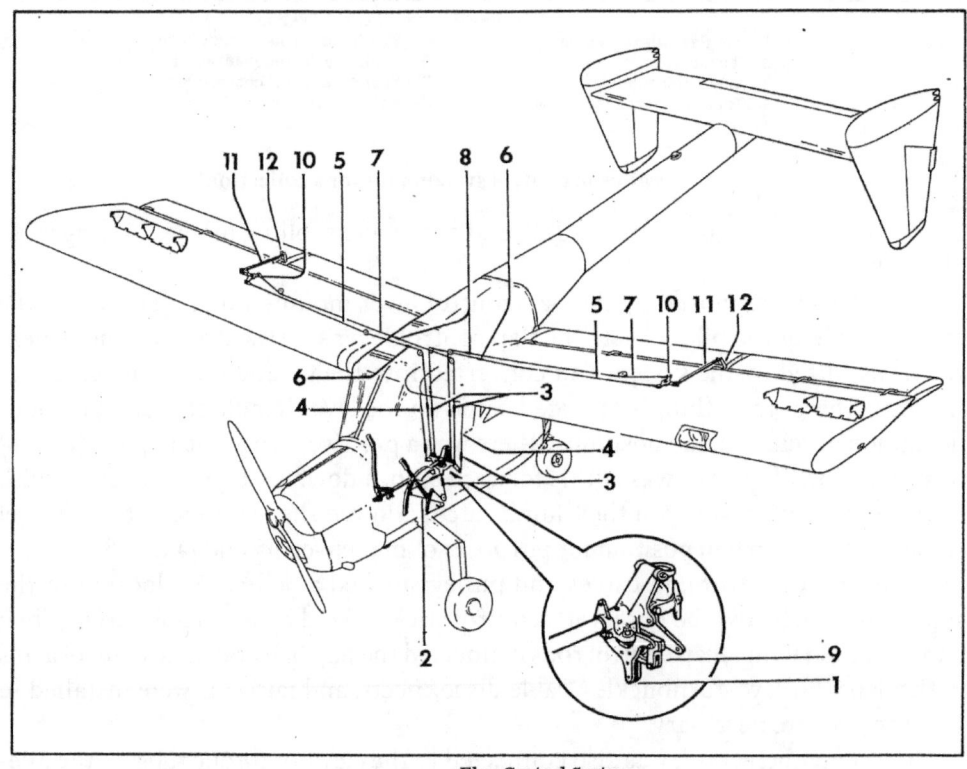

Flap Control System

1. Bellcrank
2. Flap control lever
3. Turnbuckles
4. Spoiler aileron cable
5. Flap up cable
6. Spring disconnects
7. Flap down cable
8. Loop disconnects
9. Flap control quadrant
10. Flap bellcrank
11. Flap control rod
12. Flap control arm

Flap control system (author's collection).

There was a static port on each side of the fuselage approximately in line with the instrument panel. Static ports were installed flush with the fuselage skin paneling on each side of the airplane above and slightly forward of the landing gear sponson tube. Static vent tubing lines ran from the ports to be joined together in a tee fitting which attached to a four-way fitting behind the instrument panel. Individual lines led to the static ports on the airspeed indicator, rate of climb indicator, and altimeter from this fitting. The joining of the two static port lines physically adjusted for any yawing of the aircraft. The tubing lines were accessible inside the cockpit, from below the dash, and from the rear of the hinged instrument panel.[22]

AN5812–1 pitot tube mounted below the left wing (author's collection).

Instruments

There were six flight instruments, five engine instruments, a clock, a standby compass, and a thermometer installed in the Boeing YL-15. Three of the flight instruments were gyroscopic, three were pressure-operated, and three of the engine instruments were electrical. The flight and engine instruments were located on the instrument panel's hinged section, and access to them was gained by folding down the hinged instrument panel. The standby compass was mounted directly above the panel, and the free air thermometer extended through the transparent plastic windshield panel in the upper-right-hand corner of the cockpit.[23]

The type B-8A airspeed indicator was on the upper left section of the instrument panel. It was hooked to the total and static lines of the pitot-static system using the difference between air pressure at the pitot tube and air pressure at the static vents to determine the airplane's indicated airspeed, shown on the dial in miles per hour. The airspeed indicator dial was marked at 200 mph maximum allowable speed, 112 mph max maneuvering speed with flaps less than 10° down, and 80 mph max speed with flaps over 10° down. The *Pilot's Handbook* had a correction table for Indicated to Calibrated Air Speed at different flap settings.[24]

An AN5825–3 rate of climb indicator was in the center of the lower row of

1. Light switch panel
2. Clock
3. Airspeed indicator
4. Standby compass
5. Radio call plate
6. Directional gyro indicator
7. Artificial horizon indicator
8. Voltmeter
9. Rate of climb indicator
10. Turn and bank indicator
11. Altimeter
12. Ammeter
13. Engine gage
14. Cover plate
15. Tachometer
16. Fuel gage
17. Fuel warning light
18. Instrument panel hinge
19. Manual primer
20. Cabin heat control
21. Oil gage filler check valve
22. Parking brake control
23. Starter switch
24. Ignition switch
25. Fuel selector valve indicator
26. Carburetor air heat control

YL-15 instrument panel (author's collection).

instruments on the panel. It was actuated by registering the changing air pressure at the static vents indicating the rate of change of altitude in feet per minute.[25]

An AN-GG-A-461A altimeter was located on the lower-left section of the instrument panel. It was a sensitive atmospheric pressure indicator hooked to the static ports. The dial showed the altitude in feet and had three pointers indicating 1 foot, 100 feet, and 1000 feet. There was an adjusting knob located on the lower-left corner of the case to set the local barometric pressure indicated in the small window on the right-hand side of the dial.[26]

The YL-15 was equipped with an AAF Specification No. 94-27393A type C-1 directional gyro indicator which established, by means of an electrically driven gyro, a reliable flight reference for directional (azimuth) control of the airplane. Indications were shown on the face of the instrument by a vertical compass card or dial, which was read in relation to the lubber (center) line. This made it possible to determine at a glance the heading to be flown when making 45°, 90°, or 180° turns.

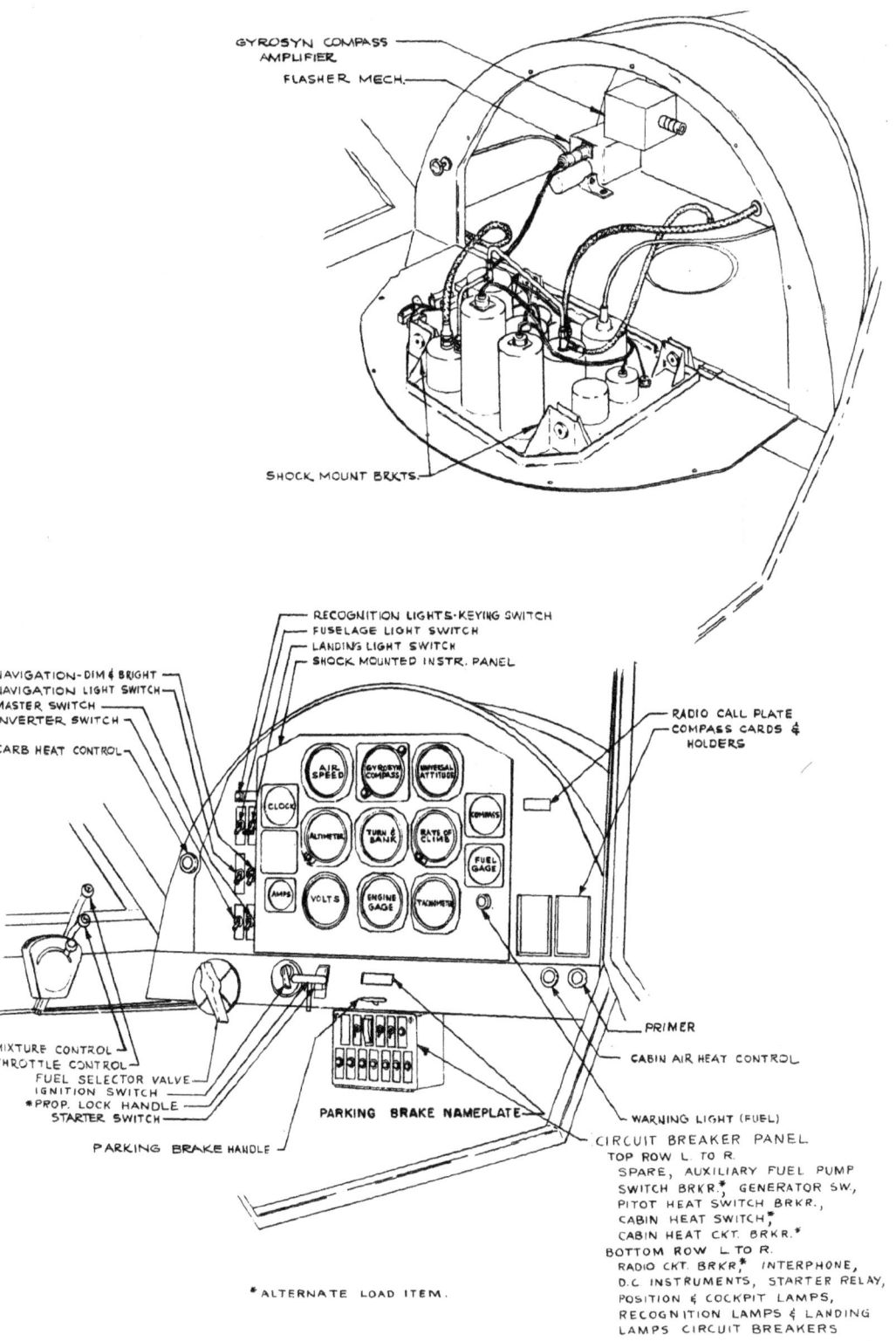

XL-15 fold-down instrument panel. YL-15 differs only in instrument layout (NARA).

AIRSPEED INSTALLATION CORRECTION TABLE				
	CORRECTION AT VARIOUS FLAP SETTINGS			
I.A.S. (M.P.H.)	10° UP	0°	20° DOWN	35° DOWN
20		----	----	ADD 15 M.P.H.
30		----	ADD 10 M.P.H.	ADD 11 M.P.H.
40		ADD 9 M.P.H.	ADD 8 M.P.H.	ADD 8 M.P.H.
50		ADD 6 M.P.H	ADD 6 M.P.H.	ADD 5 M.P.H.
60		ADD 5 M.P.H.	ADD 4 M.P.H.	ADD 2 M.P.H.
70		ADD 4 M.P.H.	ADD 2 M.P.H.	----
80		ADD 3 M.P.H.	----	----
90		ADD 2 M.P.H.	----	----
100		ADD 1 M.P.H		

NOTE: Above data based on flight tests 8 March 1948.

Figure A-1. Airspeed Installation Correction Table

Pilot's handbook airspeed correction table (author's collection).

A knob was provided to cage the gyro and also set the card at the desired heading. Turning on the airplane power supply starts the gyro. The indicator was in the upper-left section of the instrument panel.[27]

Located in the upper-right section of the instrument panel was a type E-1 artificial horizon indicator which established a reliable flight reference for control of the airplane in roll and pitch using an electrically driven gyro. The artificial horizon indicator consisted of a movable horizon bar and bank pointer, with a fixed miniature airplane in the center of the dial. A caging knob marked "PULL TO TURN" and a small knob to adjust the position of the miniature airplane were located below the dial. The attitude of the airplane in roll was indicated by displacement of the bank pointer and by the tilting of the horizon bar from a level position. Motions of the airplane in pitch were shown by the vertical displacement of the horizon bar. The relation of both bank pointer and horizon bar to the miniature airplane showed the airplane attitude in both pitch and roll. The instrument could be caged and made non-indicating by pulling out on the caging knob, turning it clockwise as far as it would go, and then pushing the knob in. It was uncaged in the same manner by turning the knob counterclockwise. When due to loading or other conditions the airplane was flown "nose up" or "nose down" for level flight, the miniature airplane would remain above or below the horizon bar while the airplane was flying level, but this could be adjusted by turning the small knob until the miniature airplane was lined up with the horizon bar during level flight.[28]

An AN5825–3 type C-1 turn and bank indicator was located in the left-hand side of the bottom row of the instrument panel. It consisted of a curved spirit level with an inclinometer ball in a glass tube and an electrically operated gyro unit. The spirit level, or bank indicator, showed the airplane deflection off transverse level attitude and the gyro turn indicator showed the rate of turn by the pointer movement.[29]

An AN5773-1A engine gauge was located on the lower right section of the instrument panel. It was a combination oil temperature and oil pressure indicator. The oil temperature was indicated in degrees centigrade and the oil pressure in pounds per square inch. The oil temperature dial was marked with "88°C min" and "105°C max." The oil pressure dial was marked with "60 psi min" and "80 psi max."[30]

There was an oil gauge filler check valve, consisting of a small filler neck and check valve with a screw cap, located at the extreme right of the instrument panel. An oil tubing line went from this filler check valve to the engine gauge unit, allowing for the introduction of low viscosity oil to the gauge to ensure correct readings during low-temperature operation.[31]

A type B-1 voltmeter was on the lower right section of the instrument panel. Either the voltage output of the engine-driven generator or the battery voltage could be read on this instrument. A conventional three-position toggle switch was installed on the circuit breaker panel to control the voltmeter readings. The switch positions were "BUSS," "GEN," and "OFF." This voltmeter switch connects the voltmeter with the generator or with the battery. The pilot could read generator voltage when the engine was running or battery voltage when the engine was inoperative.[32]

An AAF Specification No. 32529, type J-1 ammeter was located on the lower right-hand section of the instrument panel. It registered the generator load between –10 and +10 amperes.[33]

On the upper-right section of the instrument panel was an AAF Specification No. 94-27353A tachometer connected directly to the engine by a flexible shaft and indicating engine revolutions per minute on the dial. The dial was marked at 2450 rpm for max cruise and 3120 rpm for max dive rpm.[34]

A type A-11, AN5743-1A 8-day, stem wound aircraft clock was on the right-hand section of the instrument panel. The winding knob was on the lower left-hand corner of the case, and a regulating screw was accessible from the back of the clock. The clock had a sweep second hand, and the dial and hands were fluorescent coated for night operations.[35]

Mounted on a bracket above the instrument panel was a type B-21 magnetic compass.[36]

The probe for the type C-13 free air thermometer in the upper-right corner of the cockpit extended through the windshield to measure outside air temperature with the dial inside, facing the pilot. The outside air temperature dial was marked in degrees Fahrenheit.[37]

The XL-15, in general, had the same types of instruments as the YL-15, but the XL-15 panel layout was slightly different from that of the YL-15.

Electrical

The YL-15 had a 24- to 28-volt, DC, single-wire, grounded electrical system that furnished power for engine operation, lights, communicating equipment, and instruments. The system consisted of a storage battery, engine generator, inverter, carbon-pile voltage regulator, and the necessary wiring, switches, and meters. There was a receptacle for plugging in an external power source for ground operation.[38]

An Exide 12-AC-7D 24-volt, 11-ampere-hour wet cell storage battery supplied direct current to the electrical components of the airplane. The battery was located in a well below the cockpit floor, directly aft of the pilot's rudder pedals. It

XL-15 instrument panel (author's collection).

10. Systems 165

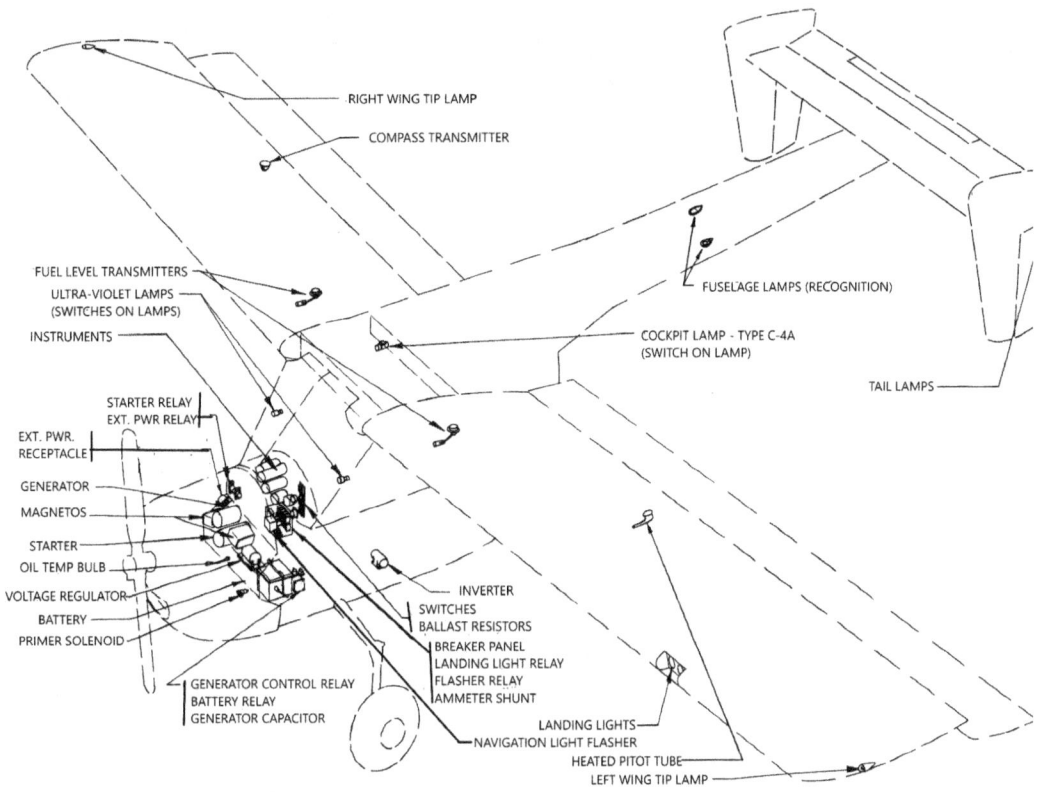

Electrical system (NARA).

was protected and enclosed by a battery access cover. The battery well was opened by unsnapping four fasteners, which attached the signal flare container to the access cover, then removing the cover after taking out four screws that secured it to the cockpit floor. The battery was controlled by an AN3022–2 two-position toggle switch on the circuit breaker panel, which wired in between the battery and all electrical circuits, functioned as a master cut-off switch for all electrical components of the airplane, exclusive of the ignition system. The battery switch was marked "BATTERY" and "OFF."[39]

The generator used was a standard-type M-3 built to AAF Specification No. 95–32361 with a rating of 50 amperes at 25.5 volts and 2500 to 4500 rpm. This generator supplied electrical power for operating the electrical equipment in the airplane and keeping the battery fully charged. A conventional carbon-pile voltage regulator regulated it. The generator was engine driven and mounted at the rear of the engine.[40]

A generator switch was installed across the generator circuit so the pilot could cut out the generator if its operation caused undue radio interference. The switch was an AN3022–2 momentary-position toggle switch located on the circuit protector panel beneath the instrument panel. A red plastic guard over the switch held it closed when the guard was closed and allowed it to be opened when the guard was pulled up.[41]

A type 24950 voltage regulator was used to maintain a constant generator output between 24 and 28.5 volts. The regulation was accomplished by means of a carbon-pile resistance at one end and a compression screw at the opposite end, which was attached to a movable armature. The movable armature was supported by leaf spider-shaped springs, the legs of which rested upon a bimetal combustion ring. A paralleling coil and an operating coil were enclosed in the case at one end of the carbon pile. As the current passed through these coils varied, the pressure on the carbon discs would be varied, affecting a differential control that kept the voltage constant within the predetermined limits. The voltage regulator was on the firewall inside the fuselage and could be reached by getting underneath the instrument panel.[42]

Six conventional reset push-button circuit breakers protected all engine, instrument, and light circuits on the airplane. They were located on the circuit breaker panel, and each button snapped out beyond its normal position when its circuit overloaded beyond the safety point. Pushing the button in reset the breaker and closed the circuit for protection against overload. There was one breaker for each circuit except for the navigation flasher, which had a fuse. The circuit breaker panel was located below the center of the instrument panel, just below the parking brake control.[43]

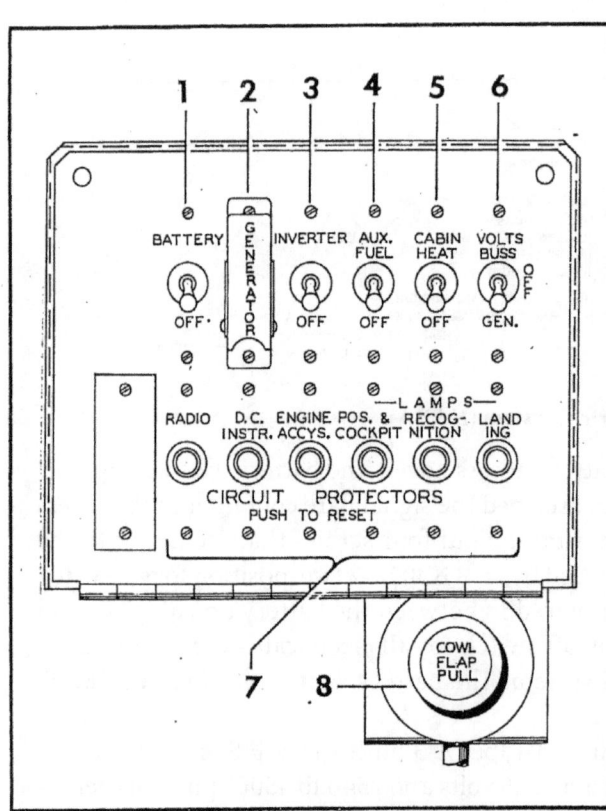

Circuit Breaker Panel

1. Battery switch
2. Generator switch
3. Inverter switch
4. Auxiliary fuel pump switch
5. Cabin heat switch
6. Voltmeter switch
7. Circuit breakers
8. Engine cowl flap control

Circuit breaker panel (author's collection).

An external power receptacle was located under a spring-closed flap door in the right-hand hinged engine cowl panel near the firewall. This receptacle was an AN2552-1 bayonet-type plug receptacle. The external power receptacle provided a means of connecting an outside power source to the electrical system for the ground operation of the airplane.[44]

A standard AN3025-1 reverse current relay was located on the forward face of the firewall near the lower edge. It was accessible by raising the hinged right-hand side engine cowl panel. The relay was automatic in operation and wired in between the generator and the battery. Usually, the reverse current relay was closed when the generator was supplying normal power to the line. When generator output dropped, as when the engine was stopped, the relay would open and prevent the battery from supplying power to the generator.[45]

Located on the pilot's cockpit floor was an AN3187–1 Class A 115-volt, 400 cycle three-phase inverter, which provided power in a two-wire AC system for the artificial horizon and the directional gyro. The inverter was controlled by a two-position toggle-type switch located on the circuit breaker panel and marked "INVERTER" and "OFF."[46]

The YL-15 had both interior and exterior lights as standard equipment. The interior lights were two instrument lights and a cockpit light. The exterior lights were navigation, recognition, and landing lights. All the light circuits were controlled by toggle switches on the light switch panel on the left-hand side of the instrument panel. The airplane battery and engine generator normally furnished power for the lights.[47]

Two ultraviolet, adjustable, type C-8 instrument panel lights were mounted above and aft of the instrument panel, on the cockpit's left- and right-hand sides. These lights were manually adjustable in their brackets and were on coiled wire, and

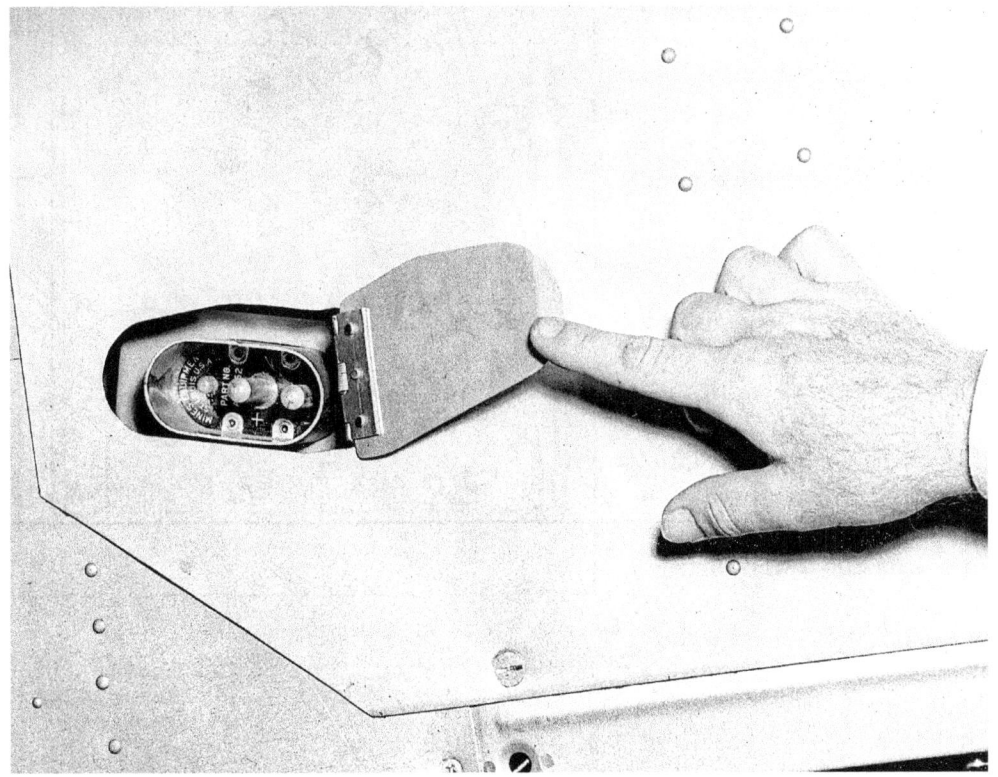

External power receptacle (author's collection).

could be easily removed from the brackets if desired and handheld. Each light socket had a push-button switch incorporated which allowed the light to be used individually. When installed in their brackets, the instrument panel lights illuminated the instrument dials during night flights. When removed from its bracket, the instrument panel light could act as a flashlight to illuminate any part of the cockpit.[48]

A cockpit light was located above the pilot's head on the right-hand side of the cable guard mounted on a bracket attached to the fuselage structure. It was identical to the instrument panel lights, except that it was a type C-4A with a white light. Its light could be directed into either the cockpit or the observer's compartment while installed in its bracket or used as a flashlight when removed from its bracket.[49]

The navigation lights on the YL-15 served to indicate the airplane's direction of travel during night flights. They consisted of a position light on each wing tip and two lights on the trailing edge of the left-hand fin. The left-hand wingtip light was red, and the right-hand light was green. Each was located in the wingtip, near the leading edge. The fin position lights were mounted one above the other in the fin trailing edge, above the left-hand rudder. The upper light was yellow, and the lower one was white. The lights were controlled by two switches on the light switch panel. The left-hand switch was a three-position toggle marked "STEADY," "OFF," and "FLASH" and the right-hand switch was a two-position toggle labeled "BRT" (i.e., bright) and "DIM."[50]

There were two recognition lights on the YL-15 airplane, one on the upper surface of the tail boom and one directly below it on the lower surface. They were located on the centerline approximately at fuselage station 250. Each recognition light unit had a large and a small white bulb. The larger light bulbs were used for signaling and the smaller light bulbs for

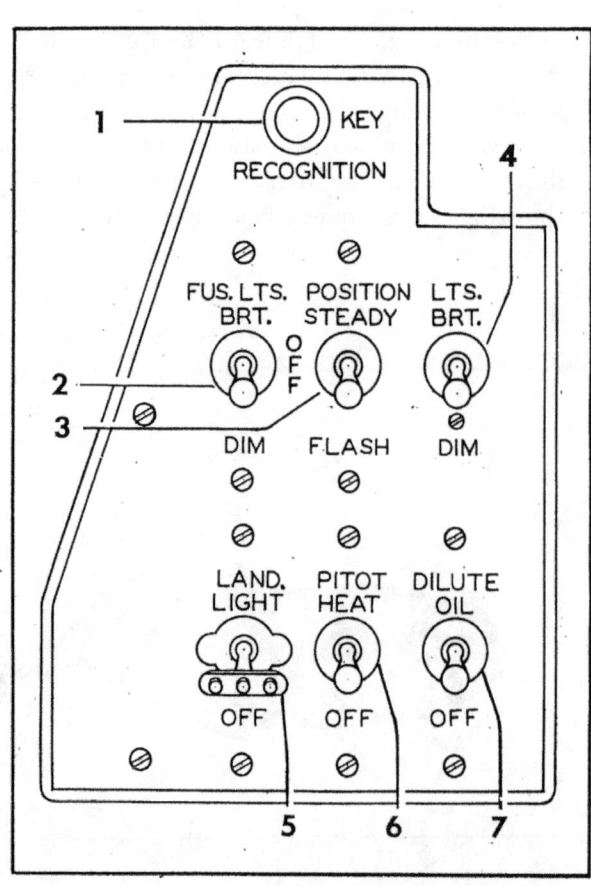

Light Switch Panel

1. Recognition key
2. Fuselage light dimmer
3. Position light switch
4. Position light dimmer
5. Landing light switch
6. Pitot heater switch
7. Oil dilution switch

Light switch panel (author's collection).

recognition purposes. The lights were controlled by the recognition key (push button) on the light switch panel. A three-position toggle labeled "FUS LTS" below the recognition key and marked "BRT,", "DIM," and "OFF" also controls these lights.[51]

Two landing or approach lights were installed side by side in the leading edge of the left-hand wing panel, approximately at wing station 200, and served to illuminate the ground ahead of the airplane for takeoff, approach, and landing. The lights were mounted side by side in a frame and were protected with a curved glass shaped to the wing leading edge. A toggle switch on the light switch panel turned these lights on and off.[52]

The reason for two adjacent landing lights in the same wing leading edge recess was the steep landing approach of the YL-15. When the airplane was sitting in its normal ground attitude, the outboard light was aimed at 3° upward from the horizontal and 2° inward, while the inboard light was aimed one-half degree downward and 8° outward. The inboard light was set for taxiing with its maximum intensity spot 155 feet forward of the airplane and 36 feet outboard of the centerline. The outboard light was set per Air Materiel Command requirements focusing on a spot 200 feet in front of the airplane, and this resulted in an extremely bright area about 100 feet from the airplane. During taxi, the contrast of this bright spot and the more dimly illuminated area beyond prevented the detection of distant objects. Thus, the second (inboard) light was added, resulting in a more diffused light covering approximately the same area. The AMC readily approved this deviation.[53]

Landing lights in the wing leading edge (author's collection).

Communications

The YL-15 was equipped for two-way voice communication with ground stations and other aircraft and the reception of radio navigation and voice signals. Dual headsets, hand microphones, and jack boxes allowed for communication by either the pilot or observer and intercommunication between the two. The equipment consisted of a complete voice amplitude modulated VHF transmitting and receiving set and a low-frequency range receiver.

Two ARC type T-11A transmitters, an ARC type R-15 receiver, an ARC type C-17 VHF receiver control unit, an ARC type C-25 VHF transmitter frequency control unit and a whip antenna provided VHF communication on six transmitting channels. The VHF equipment operated over a band of 108 to 135 megacycles, and the range receiver was tunable over a frequency range of 190 to 550 kilocycles. With normal conditions and average terrain, the VHF equipment had a range of 30 miles at 3000 feet altitude and 60 miles at 6000 feet. The VHF receiver control unit dial setting was accurate to within better than two-tenths of one percent and the range receiver control dial within one-quarter of one percent. VHF frequencies of 122.1, 122.3, 122.5, 122.7, 122.9, and 126.18 megacycles could be selected by the transmitter

Radios in the forward opening of the tail boom (author's collection).

control unit crystal selector switch, which also had an interphone position for pilot and observer intercommunication.⁵⁴

Low-frequency range and voice signal reception was provided by an ARC type R-11A receiver with an ARC type C-16 receiver control unit, a remote manually controlled loop antenna, and a fixed wire fore and aft antenna of 15 feet minimum length that ran from the top of the cabin to the top of the left fin. The loop antenna was remote-controlled by an ARC type C-18 antenna control unit. Range signals could be picked up at approximately 200 miles and, operating with the loop antenna, at approximately 100 miles.⁵⁵

A rack was provided for installing SCR619 equipment for standby use. The SCR619 was a 27 to 39 MHz FM transceiver mainly used for ground vehicles.⁵⁶

Two VHF transmitters, a low frequency receiver, and a high frequency receiver were installed in the forward opening of the tail boom. These were accessible from the observer's compartment. All operational units were shock mounted on sliding racks held in position with spring catches and could be removed without disturbing other airplane components or equipment.⁵⁷

Four control units and a jack box were located in the forward cockpit as follows: the VHF transmitter control unit was mounted above and forward on the right side of the pilot's cockpit above the door; the directional loop control unit was mounted

Radio control units above the pilot's door (author's collection).

Mast for the wire antenna that runs to the left fin and the loop antenna (author's collection).

next to and aft of the VHF transmitter control unit; the range receiver control unit was mounted slightly below and aft of the loop control unit; the VHF receiver control unit was aft of the other three control units; and the jack box was mounted forward of the VHF transmitter control unit. All were within reach of the pilot when he was in his seat.[58]

The pilot's hand microphone was hung on a hook above the pilot's seat to his left on the SCR619 radio mounting bracket. A headset was stowed aft and above the pilot's seat. The headset and microphone for the aft cockpit were stowed on cross members above the observer's seat and were within reach of the observer when he was in his seat.[59]

Antennae for the VHF equipment were mounted on the upper surface of the tail boom over the transmitters and receivers. The whip antenna was placed forward of the loop antenna. An antenna mast extended upward from the left-hand wing to the body gap and supported the wire antenna, the aft end of which was secured to the left-hand fin.[60]

The XL-15s had different radio installations than the YL-15s had. XL-15 sn 46–520 had a SCR-274N transceiver with a fixed wire fore and aft antenna of 15 feet minimum length that ran from the top of the cabin to the top of the left fin and also a MN-20A loop antenna. XL-15 sn 46–521 had the radio installation like sn 46–520 plus an SCR-619 (BC-1335) radio and an intercom system.[61]

Heating and Ventilation

During normal flight operations in temperate climates, a muff-type heater maintained adequate heat in the pilot's cockpit and observer's compartment. A cabin heat of 1.6°C (35°F) could be maintained when the outside temperature was –23.3°C (–10°F). This system consisted of a muff-type heat exchanger installed on the right-hand engine exhaust manifold and an air duct extending from the exchanger aft to the right-hand side of the cockpit. It was connected to a second duct on the fuselage floor that extended aft into the observer's compartment. A shut-off valve was installed at the forward end of the duct and was controlled by a push-pull flexible shaft with a control handle located on the lower right-hand side of the instrument panel. Manually adjustable shutters on the air duct in each compartment gave additional heat control.[62]

For colder outside temperatures, a gasoline combustion heater was installed in the lower left-hand portion of the engine compartment. This heater was a Stewart-Warner No. 977A (Mod.) electrically driven heater. It took in hot air from

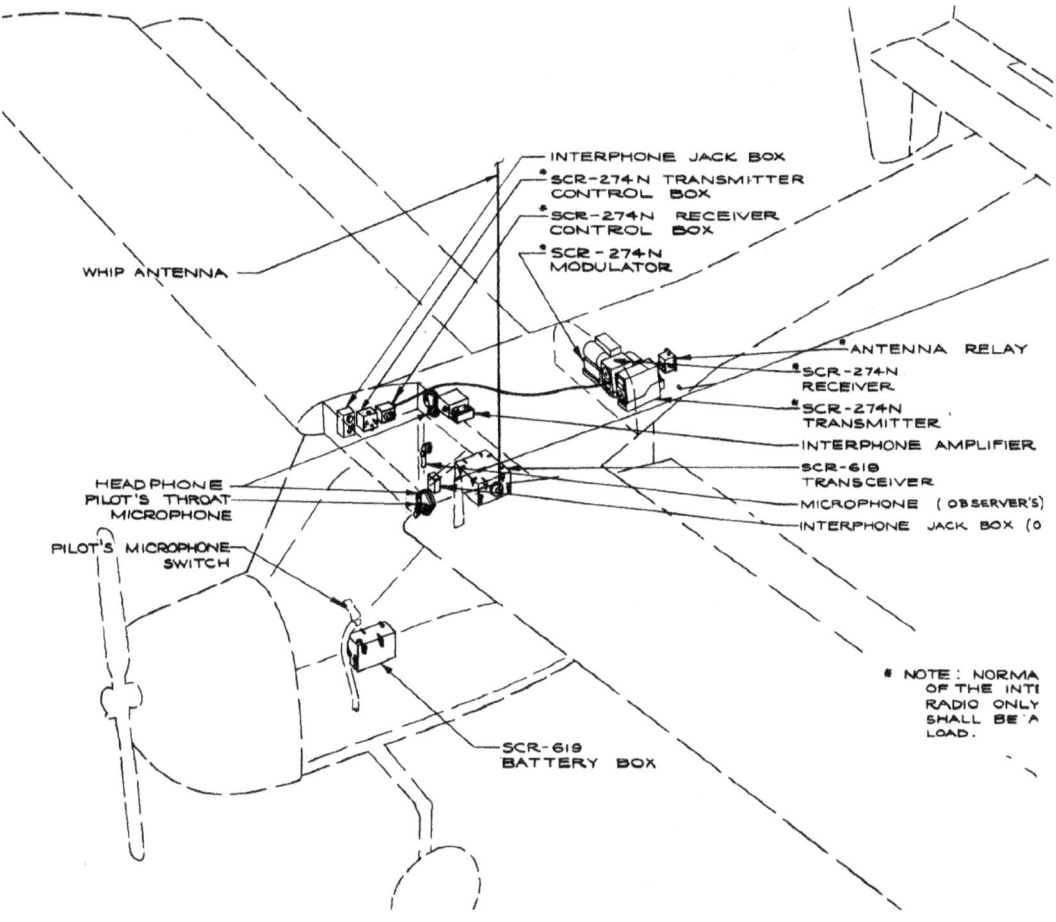

XL-15 radio equipment (NARA).

Hinged windows in the cockpit and observer's compartment provided ventilation (author's collection).

the heat exchanger on the left-hand exhaust manifold and, by the combustion of gasoline drawn from the primer line, further heated this air, which was then blown through ducts into the cockpit and observer's compartment. A cabin heat of -6°C (21°F) could be maintained with this heater when the outside temperature was -54°C (-65°F). A toggle switch located on the circuit breaker panel controlled the combustion heater. Shuttered vents for the heated air opened into the fuselage on the left-hand side of the pilot's and observer's seats.[63]

Sliding windows provided cabin ventilation. The pilot's compartment had a window on the left side which hinged at the bottom and could be opened outward and fastened in the fully open position. There was also a sliding window in the pilot's door. A triangular window on either side of the observer's compartment provided ventilation for the observer. These triangular windows were hinged at the top and opened inward and fastened by a clip on the top of the fuselage. The most effective method of cooling the pilot's cockpit was to open the observer's left window and the pilot's sliding window approximately four inches. Located in the wing leading edge at the upper outside corners of the windshield were ventilating scoops. These scoops could be adjusted by rotating their housing.[64]

11

Weights

The weight and balance of an airplane are some of its most important design attributes. They affect its performance, stability, controllability, usability, and cost. This is true no matter what type or size an airplane is.

Birth of the Liaison Airplane

The liaison airplane came about because of the failure of the observation airplane concept, primarily because of its weight. The Air Corps observation airplane had grown from a weight of 4,100 pounds in 1925 (Curtiss O-1) to 6,100 pounds in 1935 (Douglas O-46) and then to 8,100 pounds in 1939 (North American O-47). From day one the U.S. Army Air Corps observation airplanes were too heavy to operate close alongside the Army ground troops they were supposed to support and required prepared airfields well behind the front lines. By 1939–40, the Army Air Corps was under considerable criticism from the Army Ground Forces for their lack of ability to provide observation support to the artillery, infantry, and armor. The Air Corps came out with the Stinson O-49 (later redesignated the L-1) for this task, but it was still heavy and expensive with a gross weight of 3,400 pounds and an empty weight of 2,670 pounds.

The artillery had borrowed some two-place Piper, Aeronca, and Taylorcraft commercial light planes for evaluation in the 1941 maneuvers. These light planes had gross weights in the 1,200 lb range, or just over a third of the weight of the O-49, and they were able to operate off of fields and roads right next to the ground troops. Thus, the liaison airplane was born.

Boeing's L-15 Weight Effort

Boeing Wichita realized that minimizing weight was essential to the L-15 airplane program. In their pre–L-15 light plane design program, Boeing analyzed the weight and balance of existing personal and military liaison airplanes. They also had weight data for their own Stearman commercial and trainer airplanes as well as for the large Boeing bombers and transports.

In order to control the weight growth during the design of the airplane, definitive constructive weight control was exercised by qualified Weight Unit personnel assigned to carry out this objective. A. D. Kipfer was the Boeing Wichita Weight Unit Chief, and Onya A. Kelly was the lead weight engineer on the L-15 project. All drawings required the signature of the Weight Unit before release. When agreement on weight between the Project Department and the Weight Unit relative to the acceptability of the design could not be reached, the problem was submitted to the Chief Engineer for a decision.

The saving of 100 pounds in the gross weight of the Model XL-15 airplane was projected to result in

- A. Reduction in landing distance of 10 ft.
- B. Reduction in takeoff distance of 60 ft.
- C. Increase in maximum speed of 0.5 mph.
- D. Increase in initial rate of climb of 60 fpm
- E. Increase in service ceiling of 80 ft.

The Project Department adhered to the policy that no weight should be added to one part of the airplane without a corresponding reduction in weight of some other part. The success or failure of the design hinged to a large extent on the final weight; therefore, all departments concerned were asked to make every effort to save weight. Calculated weights were indicated on all drawings of assemblies, major parts, and detail parts. Once parts were produced, actual weights were tabulated on major assembly drawings such as wing panels, tail surfaces, landing gear, etc.[1]

Despite an established weight control program, the aircraft saw weight increase during its development. The proposal empty weight was 1,509 pounds, while the empty weights of the two XL-15 prototypes were 1,567 pounds for sn 46–520[2] and 1,559 pounds for sn 46–521.[3] But it must be noted that the proposed airplane had a 200-square-foot wing with a fowler flap and a single vertical tail,[4] while the XL-15s had a 268.6-square-foot wing with an external flap and a twin vertical tail. Without a good weight control program, the weight would have likely grown much more.

A slight bit of weight was trimmed out of the YL-15s, which had an empty weight of 1,557 pounds.[5]

Center of Gravity Envelope and Range

The weight control also had to ensure the aircraft had a CG range that provided for acceptable stability and control characteristics, as well as good load ability. Load ability must account for the change in CG due to crew and equipment variability as well as for fuel and its consumption. The aircraft must also be structurally sound at all points on its CG envelope.

Table 11.1. Weight and Center of Gravity

	Boeing 451-2 Proposal				XL-15 Estimate				Actual XL-15 s/n 46-521			
Model												
Reference	WD-12056, 3/6/1946				WD-10318, 9/27/1946				WD-10321, 3/5/1948			
Vertical Tail	inverted single				inverted single				inverted twin			
Flap Type	Fowler				external				external			
	Weight	Center of Gravity ~ % m.a.c.			Weight	Center of Gravity ~ % m.a.c.			Weight	Center of Gravity ~ % m.a.c.		
Load Condition	lbs	Normal	Most Fwd	Most Aft	lbs	Normal	Most Fwd	Most Aft	lbs	Normal	Most Fwd	Most Aft
Landplane Empty Weight	1509	32.0			1509.0	32.7			1559.0	37.8		
Landplane Gross Weight	2050		25.4	32.9	2050.0	38.5	31.0	39.7	2102.4	42.2	35.3	43.5
Landplane Ferry (Drop Tank)					2212.0	38.3	31.0	39.7	2275.1	42.8	35.5	44.0
Landplane Maximum Overload									2493.7	44.0		
Skiplane Empty Weight					1516.5	32.0			1580.5	37.5		
Skiplane Gross Weight					2057.5	38.0	30.3	39.0	2123.9	41.8	35.0	43.0
Seaplane Empty Weight					1725.4	34.2			1719.5	39.5		
Seaplane Gross Weight					2266.4	39.1	32.5	40.2	2262.9	43.2	37.0	44.3
Maximum C.G. Travel							28.8	43.3			33.5	47.5

Boeing engineers and workers were more used to working with large airplanes like the B-29 Superfortress seen behind the XL-15 (author's collection).

There appears to be a massive change in center of gravity position between the original proposal, which had a normal-looking CG range, and the XL-15, which had a very aft CG, but this is somewhat of an illusion. The proposal design had a wing with a Fowler flap and an MAC of 60 inches and a wing area of 200 square feet. The developed XL-15 design had a wing airfoil chord of 60 inches plus a 24-inch chord external flap, giving it an *actual* MAC of 84 inches and a wing area of 268.6 square feet. But since much of the engineering work was based on coefficients normalized using wing area and mean aerodynamic chord, Boeing decided to keep using the 200 square feet and 60 inches as the *reference* wing area and MAC. In this era before computers and word processors, to change reference dimensions would have required almost every report or document to be redone by hand.

Table 11.2. XL-15 S/N 46–521 Actual Weights from WD-10321

Item	Weight, lbs		Moment Arm, inches	
Wing Group	328.3		109.6	
Wing-Ailerons and Flaps		275.0		103.1
Ailerons		4.4		119.5
Flaps		48.9		144.8
Tail Group	83.4		301.9	
Horizontal Tail		53.3		300.8
Stabilizer		30.2		299.5
Elevator		14.0		314.8

Item	Weight, lbs		Moment Arm, inches	
Elevator Balance Weight		9.1		283.3
Vertical Tail	30.1		304.0	
Fin		17.5		301.7
Rudder		12.6		307.3
Rudder Balance Weight		0.0		0.0
Fuselage Group (Less Engine Section)	287.4		126.6	
Alighting Gear-Land Type	152.5		92.0	
Main Landing Gear	118.8		72.7	
Tail Wheel	33.7		160.0	
Engine Section	34.1		39.4	
Power Plant Group	363.2		37.4	
Engine (as installed)	255.4		33.1	
Engine Accessories	18.8		35.4	
Power Plant Controls	6.3		73.0	
Propeller	34.4		16.3	
Starting System	21.3		50.4	
Lubricating System	1.2		40.8	
Fuel System	25.8		89.8	
Fixed Equipment	306.4		93.2	
Instruments	27.6		69.7	
Surface Controls	58.5		130.1	
Hydraulic System	0.0		0.0	
Electrical	100.5		65.0	
Communications	57.3		99.5	
Furnishings	62.5		108.8	
Personal Accommodations		43.6		113.3
Emergency Accommodations		12.5		98.7
Provisions for Flight		2.3		123.9
Heating/Ventilation System		4.1		82.2
Unaccounted for Weight	3.7		221.1	
Weight Empty	1559.0		100.0	
Useful Load	543.4		110.1	
Crew (pilot n+ observer facing forward)	400.0		116.7	
Fuel (21 gallons usable)	126.0		98.8	
Fuel (trapped in system)	1.7		98.8	
Oil (2 gallons usable)	15.0		35.0	
Oil (trapped in system)	0.7		35.0	
Baggage	0.0			
Removable Equipment	0.0			
Gross Weight	2102.4		102.6	

Table 11.3 shows the weight and center of gravity of the flight crew. The observer's CG was a foot further aft when the seat was set in the forward-facing position.

Table 11.3. Crew Weights[6]

Location	Weight (lbs)	Fuselage Station (inches)
Pilot	200	92.9
Observer—Facing Aft	200	128.6
Observer—Facing Fwd.	200	140.4

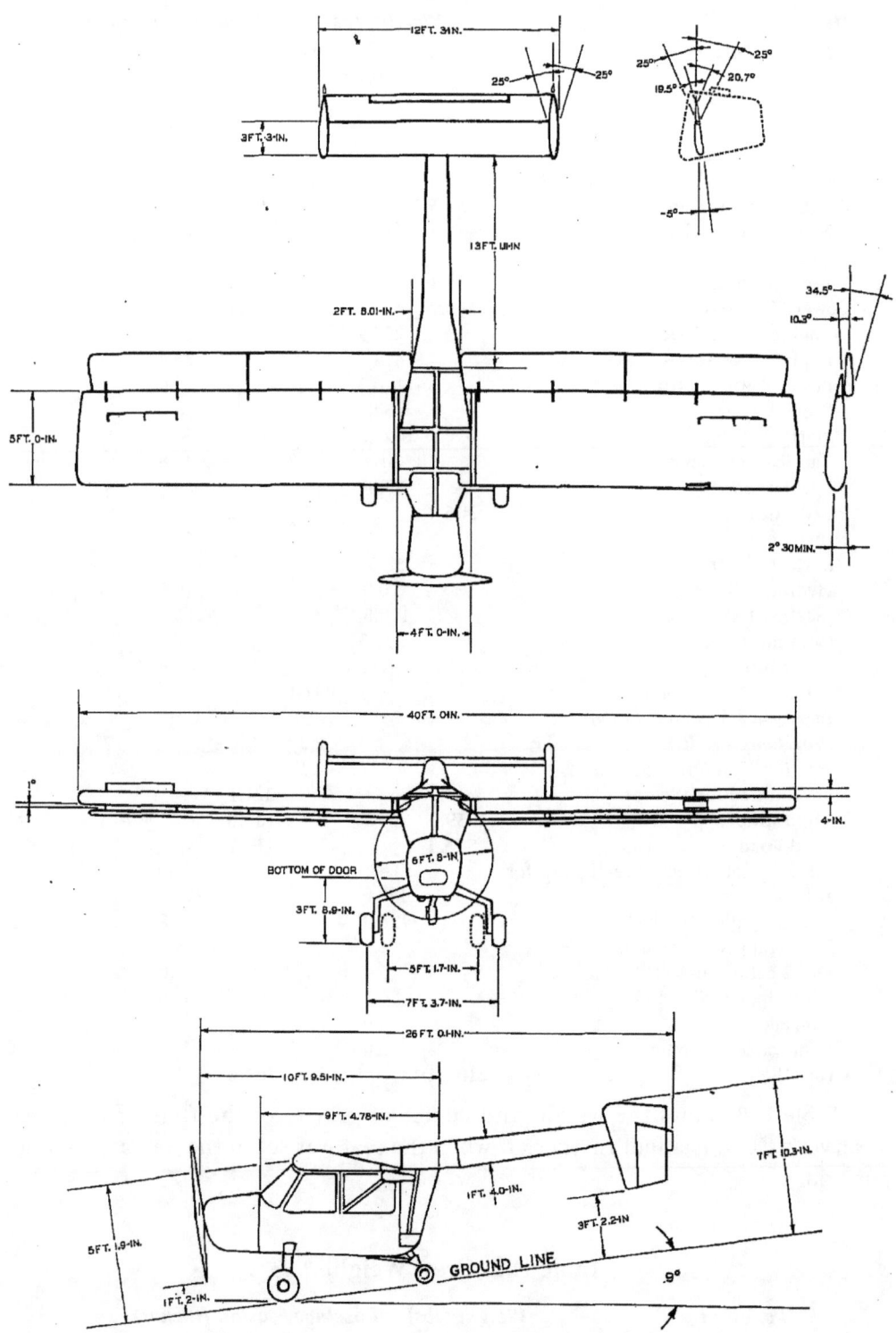

YL-15 3-view with principal dimensions (author's collection).

11. Weights

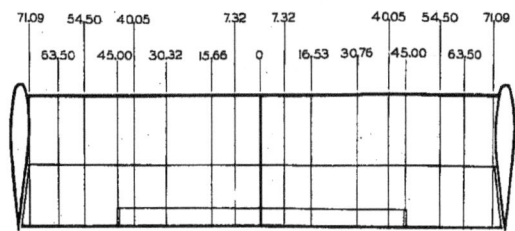

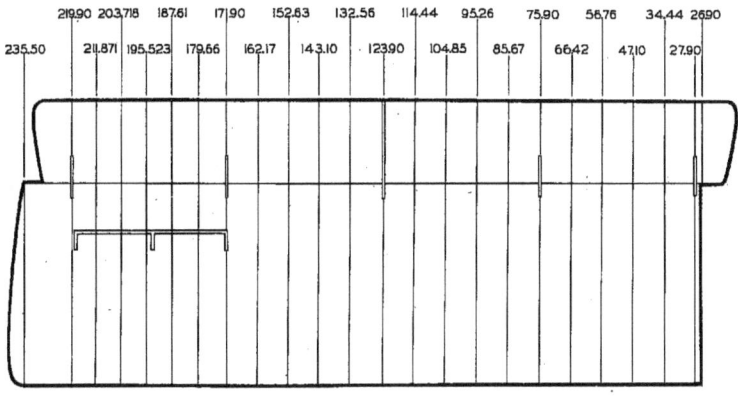

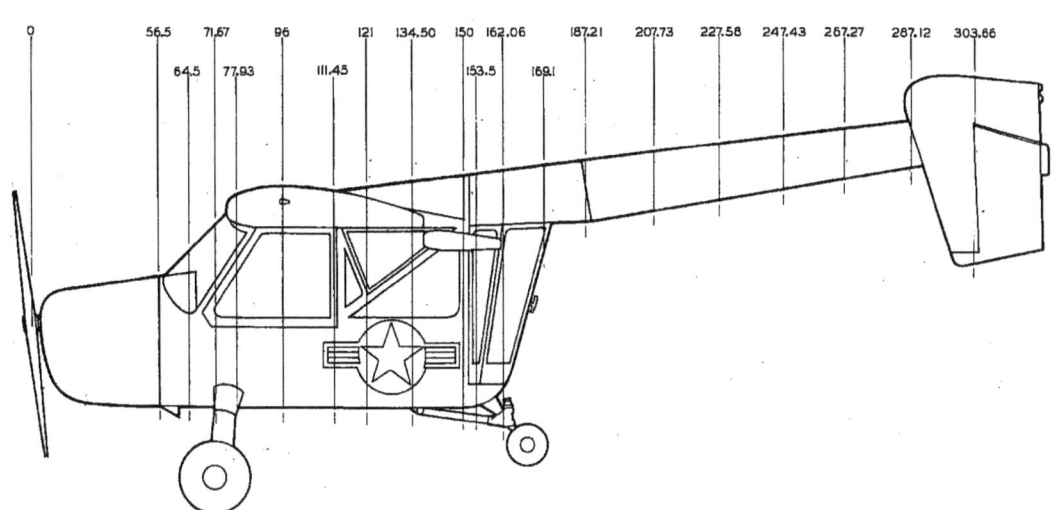

YL-15 station lines (author's collection).

Table 11.4. Optional Equipment Weights[7]

External Drop Tank (25 gal)	22.7 lbs
Winterization Package	15.7 lbs
Brodie System	31.8 lbs
Tow Release System	6.9 lbs
Propeller Lock Assembly	1.3 lbs
Message Pick-Up Equipment	1.3 lbs
Self-Sealing Tanks	39.5 lbs delta from std tanks
Armor Installation	56.4 lbs

Table 11.5. Comparative Weights of Liaison Airplanes[8]

Manufacturer	Piper	Stinson	Interstate	Convair	Boeing	Cessna
Model	L-4B	L-5	L-6	L-13A	YL-15	L-19A
Gross Weight (lbs)	1,170	2,055	1,635	2,900	2,100	2,100
Wing Group (lbs)	166.1	293.9	233.8	368.3	327.6	238
Empennage Group (lbs)	29.3	71.9	49.1	104.4	81.9	64
Fuselage Group (lbs)	134.6	258.9	202.7	394.4	291.9	216
Nacelle Group (lbs)	28.1	49.3	37.6	60.9	33.6	37
Landing Gear Group (lbs)	65.5	123.3	86.7	200.1	157.5	135
Structure Total (lbs)	*423.5*	*797.3*	*609.9*	*1128.1*	*892.5*	*690.0*
Engine (lbs)	175.5	359.6	256.7	391.5	258.3	399
Fuel System (lbs)	15	37	34.5	46.0	26.0	39
Lubricating System (lbs)	0.0	10.3	9.8	8.7	2.1	
Propeller (lbs)	15.2	28.8	26.2	75.4	33.6	46
Engine Accessories/Controls (lbs)	12.9	43.2	22.9	49.3	46.2	62
Power Plant Total (lbs)	*218.6*	*478.8*	*350.1*	*570.9*	*366.2*	*546.0*
Avionics & Instruments (lbs)	4.7	43.2	39.2	104.4	69.3	75
Surface Controls (lbs)	24.6	45.2	29.4	43.5	58.8	47
Electrical System (lbs)	0.0	84.3	36.0	92.8	102.9	86
Heating/Ventilation System (lbs)					4.1	9
Furnishings (lb)	22.2	32.9	39.2	63.8	58.9	68
Fixed Equipment Total (lbs)	*51.5*	*205.5*	*143.9*	*304.5*	*294.0*	*285.0*
Empty Weight (lbs)	693.6	1481.7	1103.8	2003.5	1552.7	1521.0
Useful Load (lbs)	476.4	573.3	531.2	896.5	547.3	579.0

12

Ground Handling and Maintenance

Jacking

Five jacking pads were provided for ground work on the airplane. One pad was located on the underside of each wing, two on the undersides of the sponson tube adjacent, respectively, to the left-hand and right-hand landing wheel, and one at the forward end of the tail wheel drag strut on the fuselage centerline. When it was necessary to work on the landing wheels on a sponson tube, the airplane could be jacked up on the wing and tail wheel pads. For work not requiring wheel or sponson removal, the airplane could be held securely on the sponson tube pads and the tail wheel pad.

When lifting the landing wheels off the ground by use of the wing jack pads, the tail wheel had to be securely tied down, and the tail boom braced against lateral movement that would cause the airplane to drop off the wing pads. The parking brake was to be locked and the landing wheels chocked before the tail wheel was raised off the ground.[1]

Leveling

Airplane leveling was accomplished by using a plumb bob and line hung from a bracket on the fuselage front spar above the pilot's head. An airplane leveling plate, on which the plumb bob's point should be centered, was fastened to the cockpit floor at the right-hand side of the pilot's seat. It had a centerline marked off in degrees and tenths of a degree. At the forward end of the centerline, a cross line was placed and marked "LEVEL." When the point of the plumb bob was directly over the intersection of the centerline and cross line on the plate, the airplane was level longitudinally and transversely.[2]

Hoisting the Airplane

There were four hoisting eyes on the fuselage, which were part of the wing upper attachment brackets. The entire airplane could be lifted by use of the airplane hoist sling, Part No. W2–150001, which consisted of a forged ring with four short wire

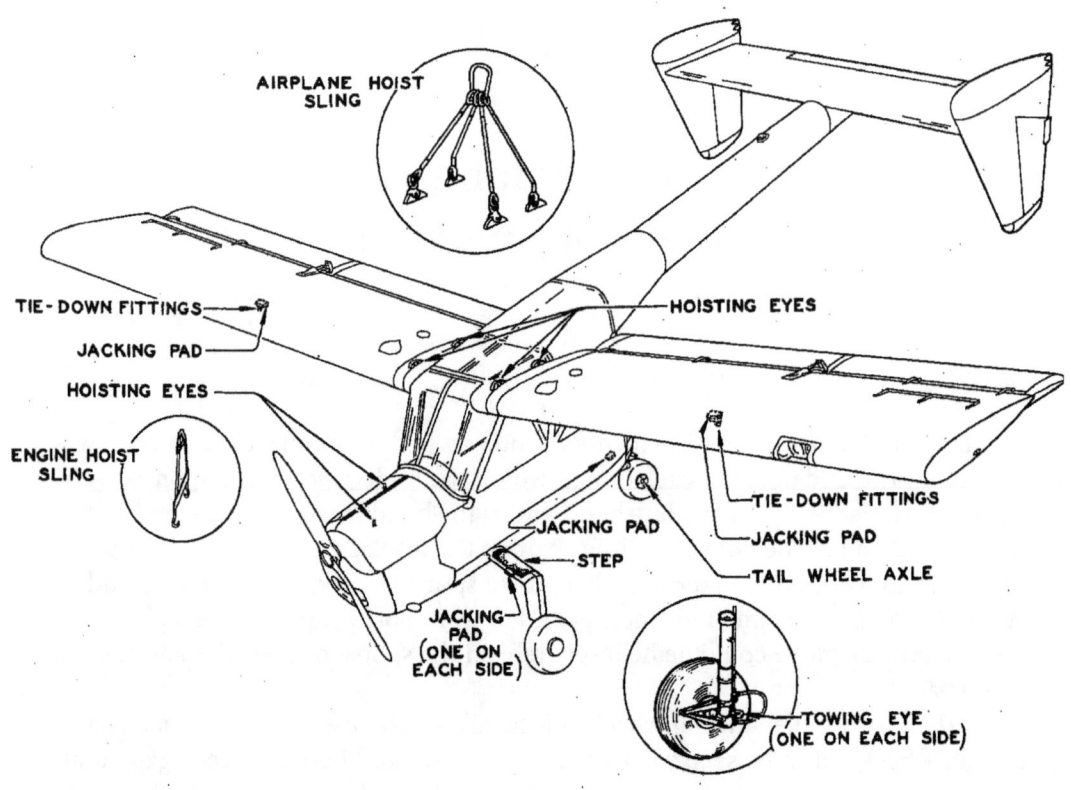

Ground handling provisions (author's collection).

cables attached. The cables had fork-end terminals, which bolted to the four hoisting eyes on the fuselage. If the airplane was to be lifted after the power plant had been removed, compensating weight must be added forward of the firewall to prevent the aft end from dropping down and damaging the fins.[3]

Hoisting the Power Plant

There was an engine hoist sling, Part No. W2–150002, which could be used to lift the engine and propeller alone. This sling had two cables that bolted to a lifting ring with forged hooks on their lower ends. There were two hoisting eyes on the engine to which the hooks of the sling were secured. The power plant was lifted just enough to keep the engine from dropping when loosening the engine mount attaching bolts, then the engine quick-change disconnects released, and the engine mount bolts removed. The power plant was now suspended on the sling, which then could be raised and moved away. Note that the engine, propeller, and accessories weighed approximately 350 pounds.[4]

The entire power plant could be removed and replaced by three men in 29 minutes.[5]

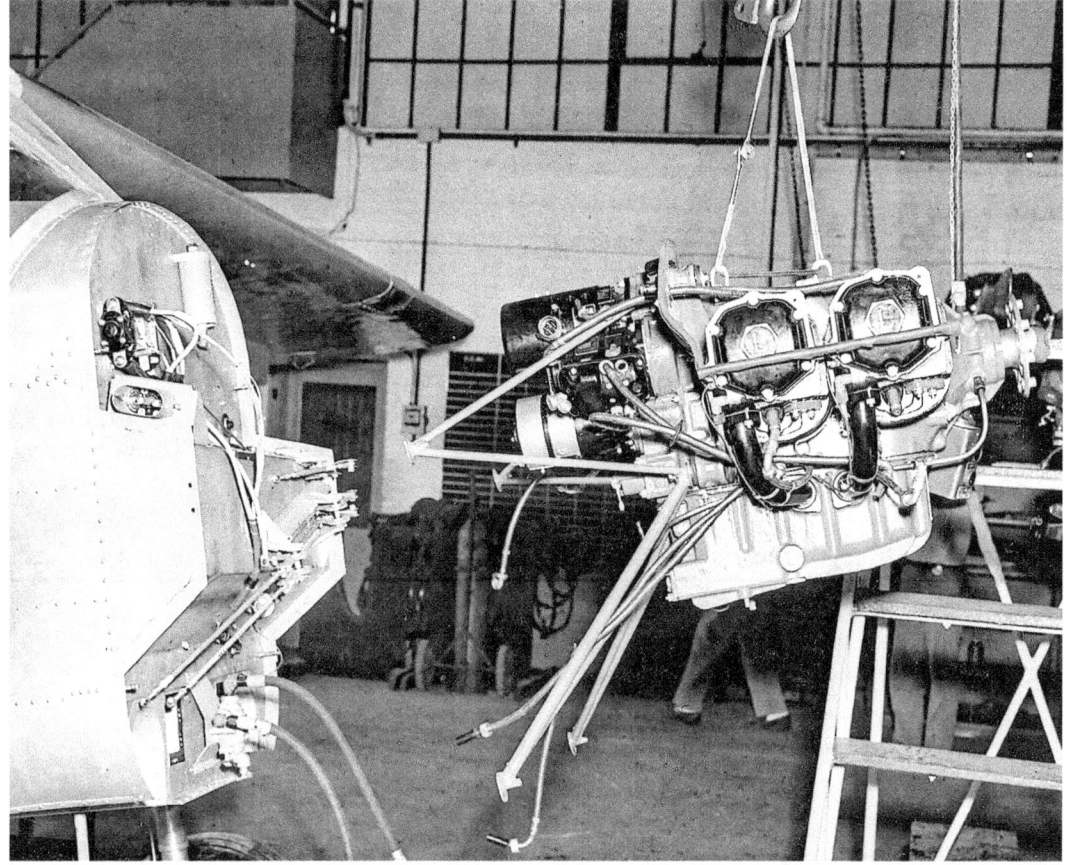

Engine hoist sling hooks to eyes on the engine (author's collection).

Parking

When the airplane was to be parked for a short time period with no adverse weather conditions anticipated, it was only necessary to set and lock the parking brake and install the parking harness. The parking harness consisted of a scissor clamp with two cables and two web straps attached and was used to lock the control surfaces when the aircraft was parked. The clamp was placed over the pilot's control stick just below the grip, and then the strap ends hooked over the observer's left and right rudder pedals. Next, the cables were fastened to the left and right keyhole fittings on the dash below the instrument panel and cables and straps were tightened by closing the lever lock on the scissor clamp. When not in use, the parking harness was stored in a bag to the left of the pilot's seat.[6]

If the brakes had been used in taxiing sufficiently to cause heating, the parking brake should not be set before allowing the brake drums to cool approximately 15 minutes. This would prevent the brake shoes from sticking to the drums.

If the airplane was to remain on the ground, not in a hangar, for a significant length of time, the aircraft should be tied down.[7]

Tie-Down

For prolonged parking, or when adverse weather was expected, the airplane should be tied down and secured according to the following procedure:

(a) The airplane headed into the wind. If gusty or violent wind was expected, the landing wheels set in shallow holes.

(b) Pass a three-quarter-inch line or one-quaerter-inch steel cable through the tie-down fittings under each wing and secure the ends to stakes set firmly in the ground. The stakes should be placed so that the two halves of these lines extended outboard 45° from the centerline.

(c) A line passed around each side of the shock absorber and through the towing eye on the main landing gear and secured to the ends of the stakes in the ground forward and aft of the wheels at a 45° angle. These lines should have been drawn up so as to compress each shock absorber slightly, the same amount on each side.

(d) A line wrapped twice around the tail wheel drag strut and secured the ends to stakes set in the ground approximately three feet from centerline.

If high winds were expected, additional security could be obtained by lashing 2 by 4 inch or 2 by 6 inch timbers, on edge, spanwise along the wings on the upper surfaces. These acted as spoilers and reduced buffeting due to high winds. The timbers were to be wrapped with padding to protect the wing surfaces.[8]

Weather Protection Covers

Covers made of waterproof duck cloth with tie ropes, zippers, and flannel lining for transparent plastic areas could be installed to provide weather protection when necessary.[9]

Individual right- and left-hand wing and flap and horizontal stabilizer covers were furnished, designed in accordance with AAF Specification 26757-B and AAF Drawings 44J19556 and 44J19562. These covers were secured in place with locking fasteners that were tightened and released by pull ropes.[10]

The fuselage (cockpit) cover fitted over the forward portion of the fuselage, covering all doors and windows completely. Tie ropes were provided at the forward and aft end of the fuselage. Zippers were installed at the top of the canopy over the wings, at the rear of the gondola portion of the fuselage, and along the boom aft of the wing trailing edge. Heavy webbing cords were secured to the aft portion of the fuselage cover so that the fuselage cover could be tied to the fuselage boom.[11]

The cotton duck power plant cover was designed in accordance with Army Specification 98–26751-G and fitted over the entire engine section.[12]

The propeller cover consisted of two pieces of cotton duck, each of which was pulled over a blade. Drawstrings were tied at the blade root.[13]

The pitot tube cover was a fabric sock provided with a red streamer to remind the pilot to remove the cover before takeoff.[14]

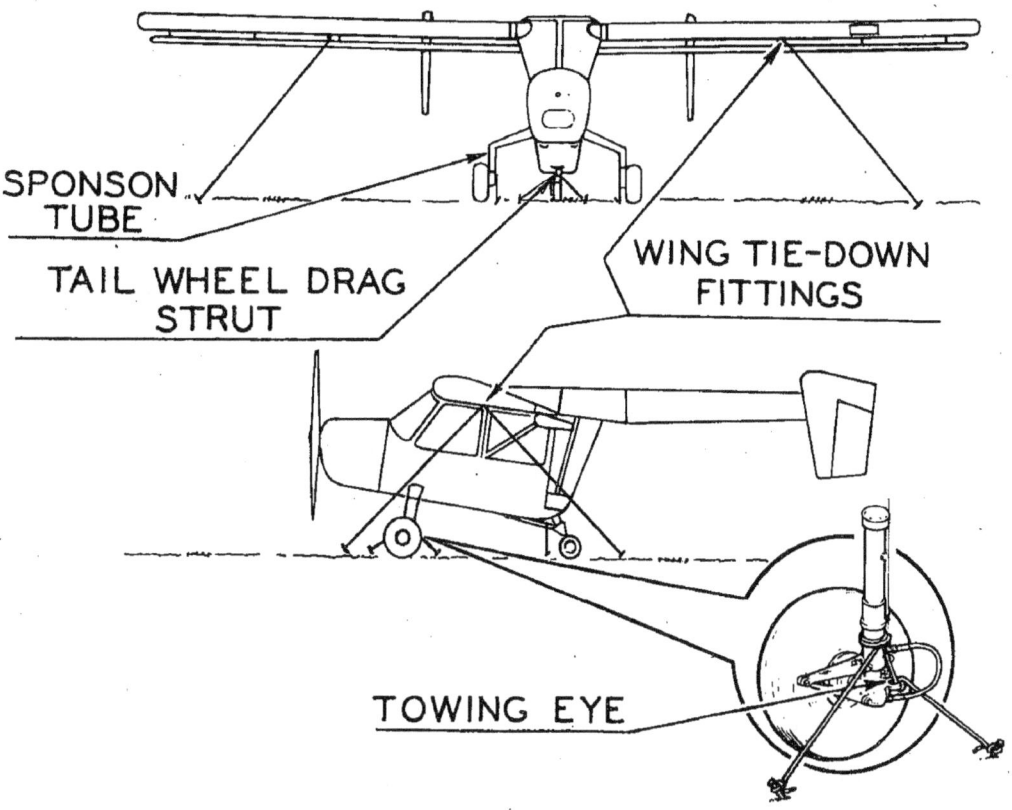

Tie-down method (author's collection).

Hand Towing

The tail wheel axle was hollow and open at both ends. This provided a means of towing the airplane by hand. A tow bar could be inserted in the axle, and the airplane could be pulled backward and steered by hand. If a tow bar was not available, any convenient pipe or rod could be placed through the axle and used to pull and steer the airplane.[15]

Ground Transportation of the Airplane

Because of the L-15's light weight, ease of assembly, and compact configuration, it could readily be loaded on a standard military two and a half ton, 6 × 6 truck or could be stowed and secured for towing behind a jeep or truck. Five men could disassemble the L-15 into a transportable package in 39 minutes or reassemble it in 45 minutes.[16]

The following procedures were done for loading or towing. During all of these procedures, three or four men were needed to handle the removed parts so they would not be damaged by dropping or sliding out of place.

XL-15 with weather protection covers on the wing, fuselage, and horizontal stabilizer (author's collection).

(a) Make the tail wheel swivelable by moving one rudder pedal full forward, then rotating the wheel beyond the rudder travel arc to disengage it from the rudder controls. Leave the tail wheel knuckle pointing forward.

(b) Install the parking harness.

(c) Remove the wings. Do not remove the flaps from the wings. Secure the wing attachment pins in their fittings on the fuselage.

(d) Remove the horizontal stabilizer. Do not remove the elevator or elevator tab.

(e) Do not remove the fins and rudders from the stabilizer, but take out the upper fin stabilizer attachment bolt from its fitting. Then fold the fins inboard, with rudders attached, so they rest flat against the stabilizer lower surface. Secure the attachment bolts back in their fittings.

(f) Rotate the landing wheels 180° to shorten the wheel tread. This was accomplished by removing the upper scissor arm attachment bolt from the lug on the lower end of each shock absorber cylinder. This lug was on the forward surface of the cylinder, and there was a similar lug directly aft of it on the opposite surface. Swivel the wheel around and secure the scissor arm to the aft lug.

(g) Attach a special loading saddle to each side of the sponson tube.

(h) Bolt the loading truss to the two wing attachment fittings on the inboard end of either wing panel. Then lift both wings up onto the sides of the fuselage, with the leading edges resting in the saddles and the inboard face of each wing toward the rear of the airplane. Extreme care must be exercised in placing the wings so as to prevent damage to fuselage transparent plastic panels. Also, be sure the protective padding was secure in the saddles.

(i) Now bolt the free wing panels to the loading truss.

(j) Lash the stabilizer and fins flat against the wing on the left side, placing the fins outboard. Use the special web straps provided and place protective padding between the stabilizer and wings.

(k) Bolt the upper left- and right-hand eyes on the truss to the two studs which extend from the sides of the tail boom above the observer's compartment doors.

(l) Disconnect the aft ends of the flap control rods from the flaps and fold the flaps down so that they lie flat, supported on the five hinge fittings on each.

(m) Tie the left- and right-hand flaps together across the top of the fuselage by bolting the ends of the special loading tie bar to each flap control arm.

(n) Secure each wing to the landing wheel below it by installing special loading web straps from the tie-down fittings on the wings to the towing eyes on the shock absorbers. The airplane will now be stowed and ready for ground transportation. Make sure that the parking brake is released before the airplane is moved.[17]

Towing by Jeep or Truck

It was recommended that a jeep or truck back up to the loading truss to couple it securely for towing. The tail boom should extend forward over the jeep or truck body. The airplane should roll backward on the landing wheels, steered from the truss coupling point, with the tail wheel swiveling freely.

For extensive ground transportation, it was recommended that the airplane be towed from the forward end by attaching the legs of the bar to the towing eyes on the landing wheels, and letting the tail wheel swivel freely. Steering was accomplished in the same manner as with any two-wheel truck trailer.[18]

Transporting by Truck

The stowed airplane was the proper width, length, and weight for transportation in a standard military two and a half ton, 6 × 6 truck, or an equivalent vehicle. The airplane was easily rolled up a ramp into the truck body and secured by adequate lashing at the sponson tube and tail wheel. It was recommended that the airplane be loaded with the aft end forward.[19]

Transporting by Cargo Aircraft

The airplane in the disassembled stowed configuration was the proper width, length, and weight for transportation in many Air Force cargo aircraft of the day

Packed up and hidden among trees (author's collection).

that had ramp loading such as the C-82 or C-97. The airplane was rolled up a ramp into the aircraft fuselage and secured by adequate lashing at the sponson tube and tail wheel. A Fairchild C-82 Packet could carry two YL-15s. The Boeing C-97 Stratofreighter could take four L-15 Scouts, 30 pilots and ground crew with equipment (at 240 pounds each), and 4,122 pounds of tools and replacement parts over a distance of 3,500 miles with a three-hour fuel reserve.[20]

This air transportability was used several times to relocate XL and YL-15s.

Maintenance and Repair

Unlike previous Army Ground Force liaison airplanes, the L-15 was not designed for repair in the field. Maintenance of the L-15 primarily consisted of keeping the fluid levels topped out, tires full of air, control cables tensioned, fuel strainers clean, and light bulbs working. The airframe was not designed for repairs, and most items were specified that if damaged they should be removed and replaced. The landing gear shock absorbers and transparent window panels were two of the few items that repair instructions were given for.[21]

Most airframe items were not even designed for repair at a depot. The L-15 was,

12. Ground Handling and Maintenance 191

Top: Towed by a jeep (author's collection). *Bottom:* The knocked down L-15 fit into an Army 2½ ton truck (author's collection).

in general, a replace the part, not repair it, airplane. If a bullet went through the wing, you did not patch the hole, you replaced the wing panel. If a stone was thrown up during takeoff or landing and damaged a flap, you did not repair the flap, you replaced it.

At the time, Army liaison sections in the field had aircraft mechanics that took care of the field service, maintenance, and repair of the liaison airplanes. Army Ground Force pilots were also trained in basic aircraft maintenance and repair. Air Force liaison squadron pilots were not trained in maintenance or repair, and Air Force liaison squadron aircraft were repaired in maintenance depots and not at the user unit level.

The airplane's lack of field reparability was likely one of the reasons the Army was disappointed in the L-15.

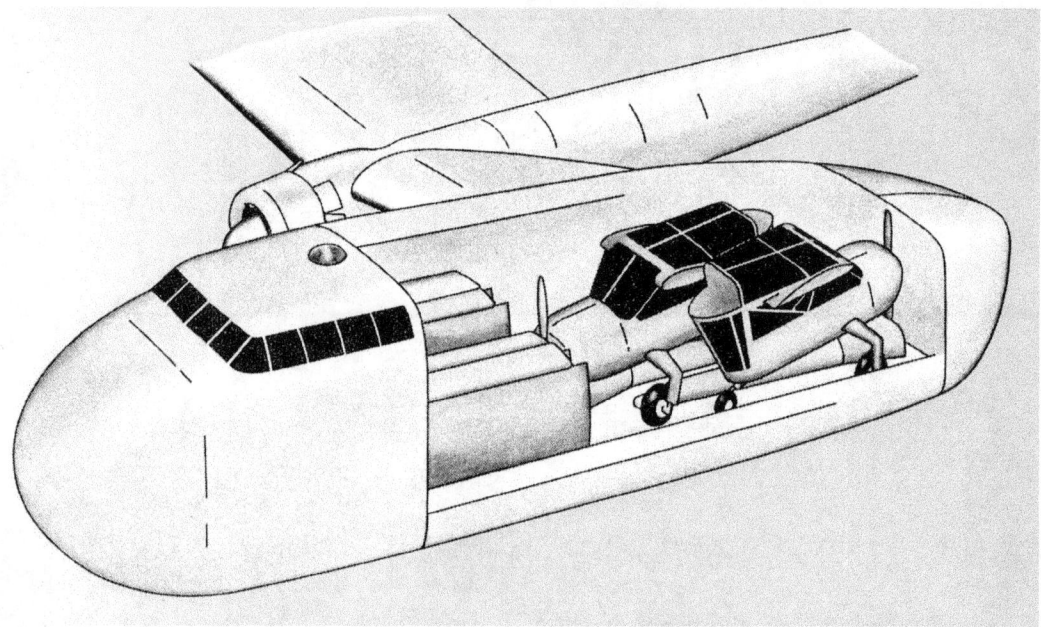

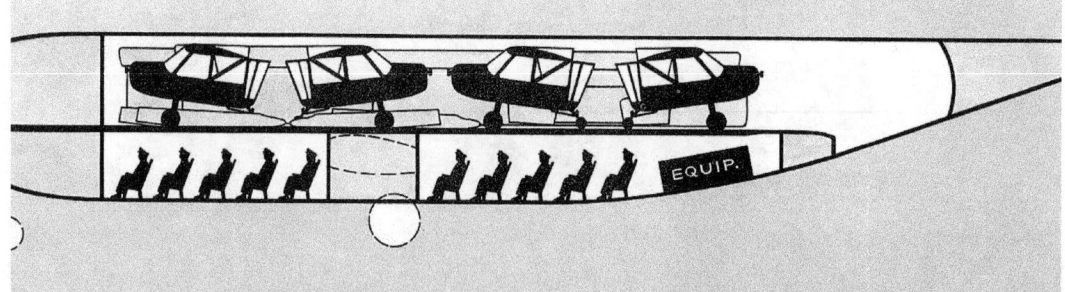

Top: The Fairchild C-82 could carry two L-15s (author's collection). *Bottom:* Boeing C-97 could carry four L-15s and their flight and ground support personnel (author's collection).

12. Ground Handling and Maintenance

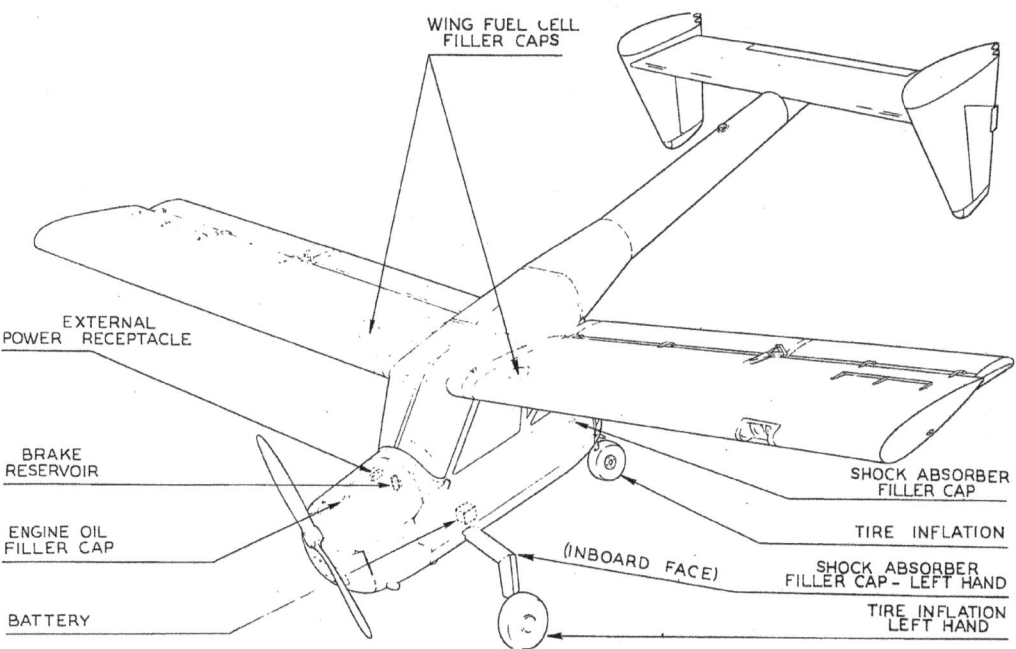

Top: XL-15 inside the cargo bay of a C-97 (author's collection). *Bottom:* Service points diagram (author's collection).

13

Boeing Pilot Comments on the YL-15

In the spring of 1948, the Air Force and Army concluded that, short of a complete redesign, there was no benefit to further modifying the XL-15 configuration. Even though the aircraft was somewhat lacking in performance, it was decided to use the available funding to order ten YL-15s for field evaluation by the Army Field Forces. The details of the order of the YL-15 are covered further in a later chapter.

After Boeing's experience with the Air Force flight test pilots' difficulty flying the two XL-15s, the company decided to prepare a briefing document on flying the YL-15 before the aircraft went out for line service. This document was Boeing Report WD-12827 "PILOT COMMENTS AND FLIGHT PERFORMANCE DATA ON THE YL-15 AIRPLANE" by D. C. Heimburger, L-15 Project Pilot, dated March 8, 1949. Douglas C. Heimburger (1920–1951) was born and grew up in China, where his father was a medical missionary. Doug had been an Army Air Forces pilot during World War II and came to Boeing Wichita Flight Test after graduating from Purdue University with a degree in Aeronautical Engineering. Among his tasks at Boeing, he was the YL-15 Project Pilot. He was also one of the first Boeing Wichita B-47 test pilots and, unfortunately, would later die on Labor Day weekend 1951 in a collision of two B-47s.[1]

This chapter is based on Boeing's briefing Report WD-12827. The document told that, late in 1945, the Boeing Wichita Engineering Department had started the development of a wing and flap combination that could produce extremely low-speed flight characteristics with acceptable lateral control. This research program had included considerable wind tunnel and flight testing and resulted in a wing configuration that gave the desired low-speed characteristics with sufficient lateral control. It was this background that the Boeing Wichita plant used in their winning bid in the XL-15 competition.

WD-12827 noted that the military in the XL-15 competition desired the 125 horsepower Lycoming 0-200-7 engine to be used. Many tough and unusual problems arose in trying to satisfy the flight and performance requirements. In fact, in this airplane's design and development, the problems encountered in the low-speed flight range were just as numerous and difficult to solve as those in the high-speed range. Doug Heimburger stated that all specifications were either met or exceeded except for the service ceiling, which was 13,400 feet instead of the specified 15,000 feet.

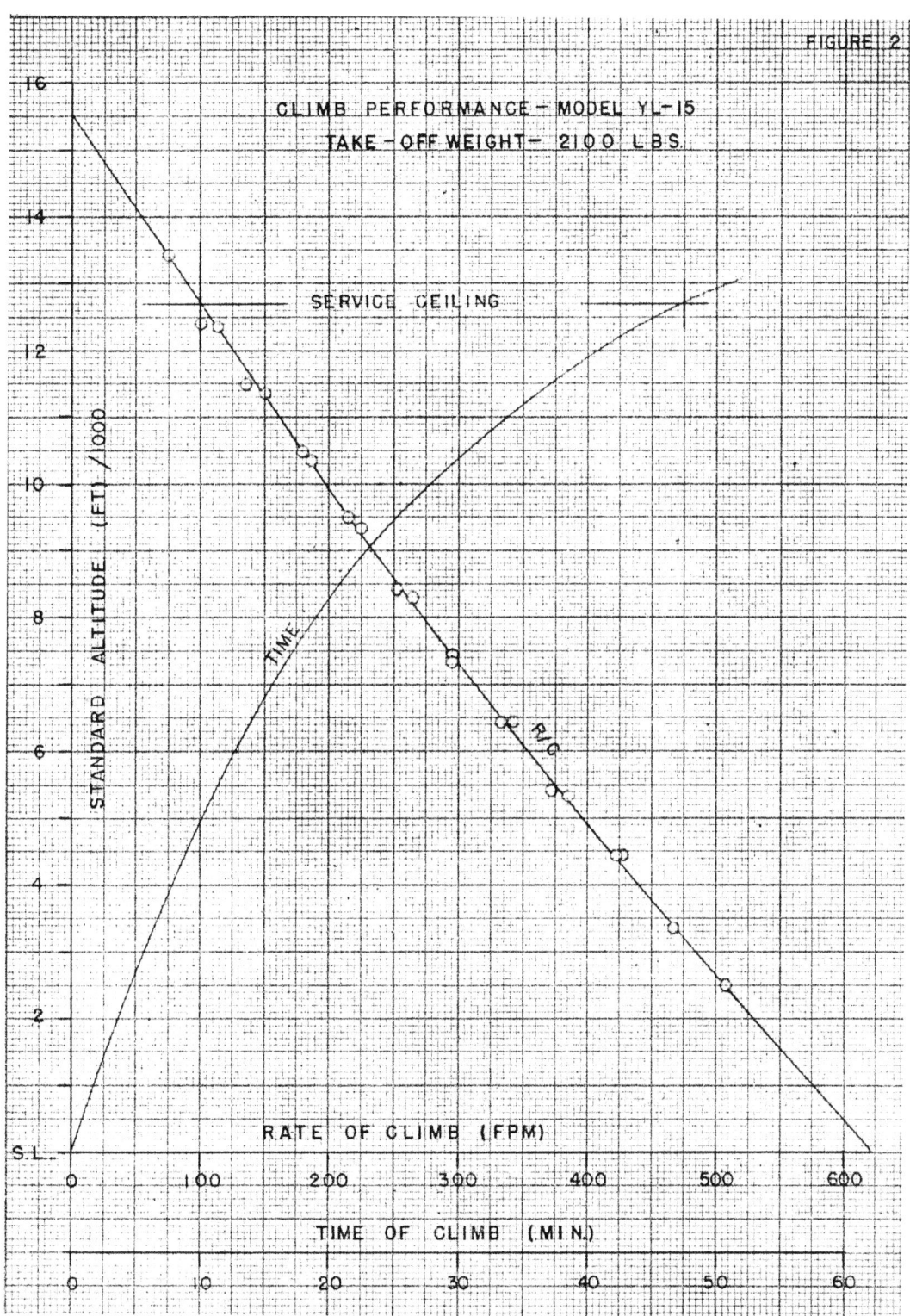

YL-15 climb performance (author's collection).

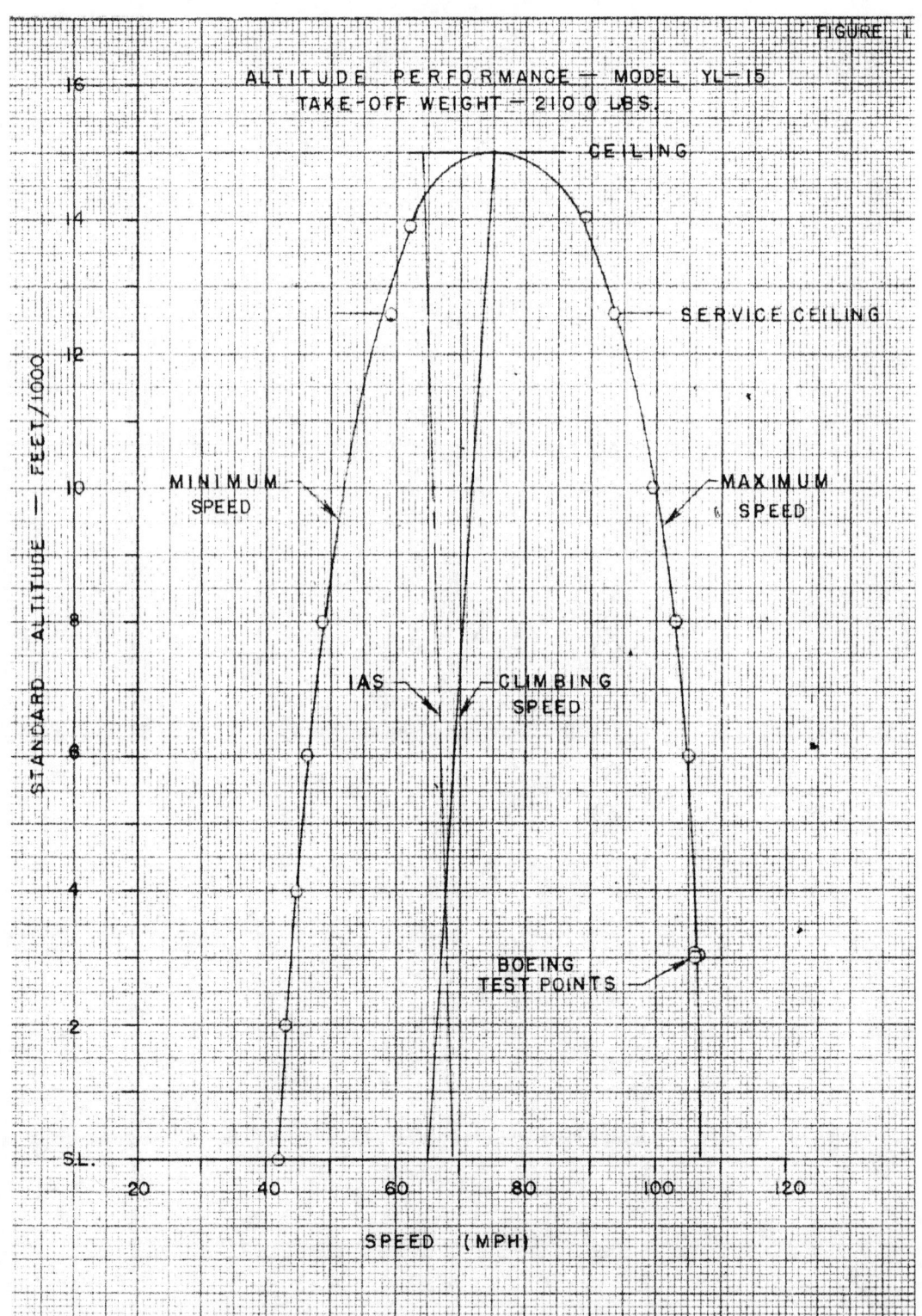

YL-15 altitude performance (author's collection).

Boeing included many unusual features in the L-15 to add to the utility of liaison-type aircraft, and it built the plane to the rigid requirements of tactical military aircraft. Boeing sincerely hoped the design features would be beneficial to the Army Ground Forces by increasing the usefulness of liaison-type aircraft and decreasing the problems confronting the pilot and crew.[2]

Special Features

Features of the L-15 that distinguished it from other airplanes of this class were its excellent visibility and the full-span large area flaps. The ratio of flap area to wing area was greater on the L-15 than on any other commonly known airplane of its day.

The large flaps made it possible to attain exceptionally high lift coefficients that gave the airplane short-field takeoff and landing characteristics. Boeing Wichita flight test data indicated that a full flap, full power lift coefficient of approximately 5.0 and a full flap, power off lift coefficient of approximately 2.7 could readily be developed, whereas Air Force flight test data indicated that a full power lift coefficient much higher than 5.0 had been attained. For comparison, most conventional liaison airplanes of that day had a maximum attainable lift coefficient of roughly 2.0.

As the flaps were moved throughout their deflection range, an extremely large range of nose attitudes in relation to the horizon was necessary to hold a constant altitude. This was advisable to keep in mind in case of a "go around" from a landing approach with flaps deflected, as the nose should be at a very low attitude to maintain climbing speed.

The YL-15 could be "jumped" vertically, approximately 125 feet at 80 mph IAS, by suddenly deflecting the flaps and holding the nose attitude relatively constant.

Doug explained the short-coupled landing gear arrangement was made necessary by the shape of the fuselage. But with the steerable tail wheel attachment, no adverse effects had been noticed, and the ground handling characteristics were superior in many ways to those of other airplanes in this class. He noted that compared to the normal tail wheel arrangement, only approximately one-half of the angular deflection of the tail wheel was necessary on the L-15 to make the same ground turning radius—this was apparent in the positive reaction of the airplane to the movement of the tail wheel. A positive locking device was included in the steering mechanism to ensure the tail wheel was kept in the steerable condition except when making very short radius turns.[3]

Taxiing

The steerable tail wheel was connected to the rudder pedals and provided positive taxi control. The tail wheel "broke" and became a full-swivel wheel after it turned more than 15°. It could be re-engaged by simply moving the airplane straight for a foot or two using the brakes and throttle, then moving the rudder pedals about their neutral position until a definite "click" was felt or heard as the tail wheel locked back in the steerable position. Full rudder travel could be used without unlocking the tail wheel as long as the wheel itself did not turn more than 15°. The tail wheel should be kept in the steerable condition except when necessary to make very short radius turns.

The whole empennage was located out of the propeller slipstream, which resulted in the rudder and elevator having only a very slight effect during taxiing; however, the tail wheel arrangement provided very good control under all normal conditions.

Taxiing with the flaps in the –10° setting had been found to definitely reduce the undesirable effects of gusty and high winds on the wings. The airplane had been taxied directly crosswind by use of rudder movement alone (with occasional use of brake to correct for effect of gusts necessary) in winds up to 20 mph, and the aircraft was found to be manageable in winds as high as 35 mph.[4]

Takeoff

The specifications for the airplane required a takeoff from a sod field over a 50-foot obstacle in 600 feet. This had been achieved in flight tests by both Air Force and Boeing Wichita pilots.

Flight test takeoff data were reduced to the weather conditions of zero wind and sea level on a "standard day," which is roughly an average of yearly temperature (59° F) and pressure (29.92 inches of mercury), and to the airplane's standard gross weight (2,090 pounds on the XL-15). Takeoffs at temperatures colder than standard would be shorter, and at temperatures above standard would be longer distances than at standard conditions. Winds and gross weights other than standard would also affect the takeoff distance.

The most consistent short takeoffs over a 50-foot obstacle were obtained by placing the flaps at 20°, holding brakes and applying full throttle, then releasing brakes and, as soon as speed was sufficient to provide elevator control, "boosting" the tail up into approximately a level flight attitude, then as the speed approached the stalling point at 20° flaps with full power, the airplane being "pulled" off by a sudden and almost full aft stick movement and climbed to the obstacle at minimum speed. The critical point was the speed at the "pull-off" and should be matched to individual technique and terrain conditions—if this speed was low, a shallower angle of climb to 50 feet resulted, but a shorter ground roll was required; if this speed was high, the ground roll was longer but the angle of climb was steeper. An exceptionally steep climb angle to the 50-foot obstacle was possible. It should be realized that a maximum effort takeoff to clear a 50-foot obstacle would result in a near stall just over the obstacle; hence, the nose must be lowered to gain climbing speed. To make a steady climb out to any given altitude, more speed should be gained either on the takeoff roll or should be allowed to build up as the airplane leaves the ground so that the airspeed for best climbing angle or best rate-of-climb could be attained early.

Takeoffs by suddenly deflecting the flaps after speed was gained on the takeoff roll, as well as other more conventional techniques, had been tried and found to present no unusual problems. Special attention in the precise holding of nose attitudes and control of airspeed paid off in gaining good takeoff performance.

The best all-around short field takeoff performance seemed to be with a flap setting of 20°.[5]

Flight

Doug Heimburger said the general flight characteristics of the YL-15 were above average. The airplane was designed to have better-than-average stability throughout its normal flight range, and it had exceptionally stable characteristics at "hovering" speeds just above the stall.

The flaps had a large effect on the nose attitude and made possible the low-speed capability of the airplane. Elevator stick forces caused by deflecting the flaps were not found to be excessive and could readily be overcome by the use of the positive acting elevator trim tab.

Stalls presented no special problem, but the stall speed range was exceptionally wide depending on flap and power settings. The lowest stall speeds, of course, occurred with full flap and full power. Care had to be exercised when raising the flaps to build up the airspeed since the YL-15 could be flown with flaps deflected at airspeeds considerably less than its 0° flap stalling speed. Normally a considerable "sink" was experienced when the flaps were suddenly raised, particularly at low airspeeds; however, this sink could be avoided by raising the flaps very slowly and allowing the airspeed to build up. It was especially advisable to keep this in mind when flying at very low altitudes.

Suddenly deflecting the flaps resulted in a near-vertical jump, which Boeing believed along with the sink resulting when the flaps were raised suddenly, might be useful in evasive maneuvers and in very low altitude obstacle clearance. Setting the flaps at different positions throughout their range would place the airplane in attitudes from nose high to nose extremely low while maintaining constant altitude, which could also be helpful in observation work, laying cable, or other miscellaneous situations.

Phase I spin tests on the first XL-15 prototype indicated that the airplane was practically non-spinnable. But later, when a reverse slot elevator, additional rudder area and travel, and more effective leading-edge spoilers were added to improve other characteristics of the airplane, these changes resulted in a spinnable configuration. Even with the revised configuration, there was no tendency for the airplane to enter a spin unless full rudder was applied during a stall. The spin rate of rotation was rather high, but during spin tests on the second XL-15 prototype, no dangerous or undesirable features were seen. Recovery was extremely positive, with several recoveries in one-eighth to one-quarter of a turn from a six-turn spin using normal reverse control techniques during the spin tests. Various control movements and sequences of movements were attempted to create a condition that would result in

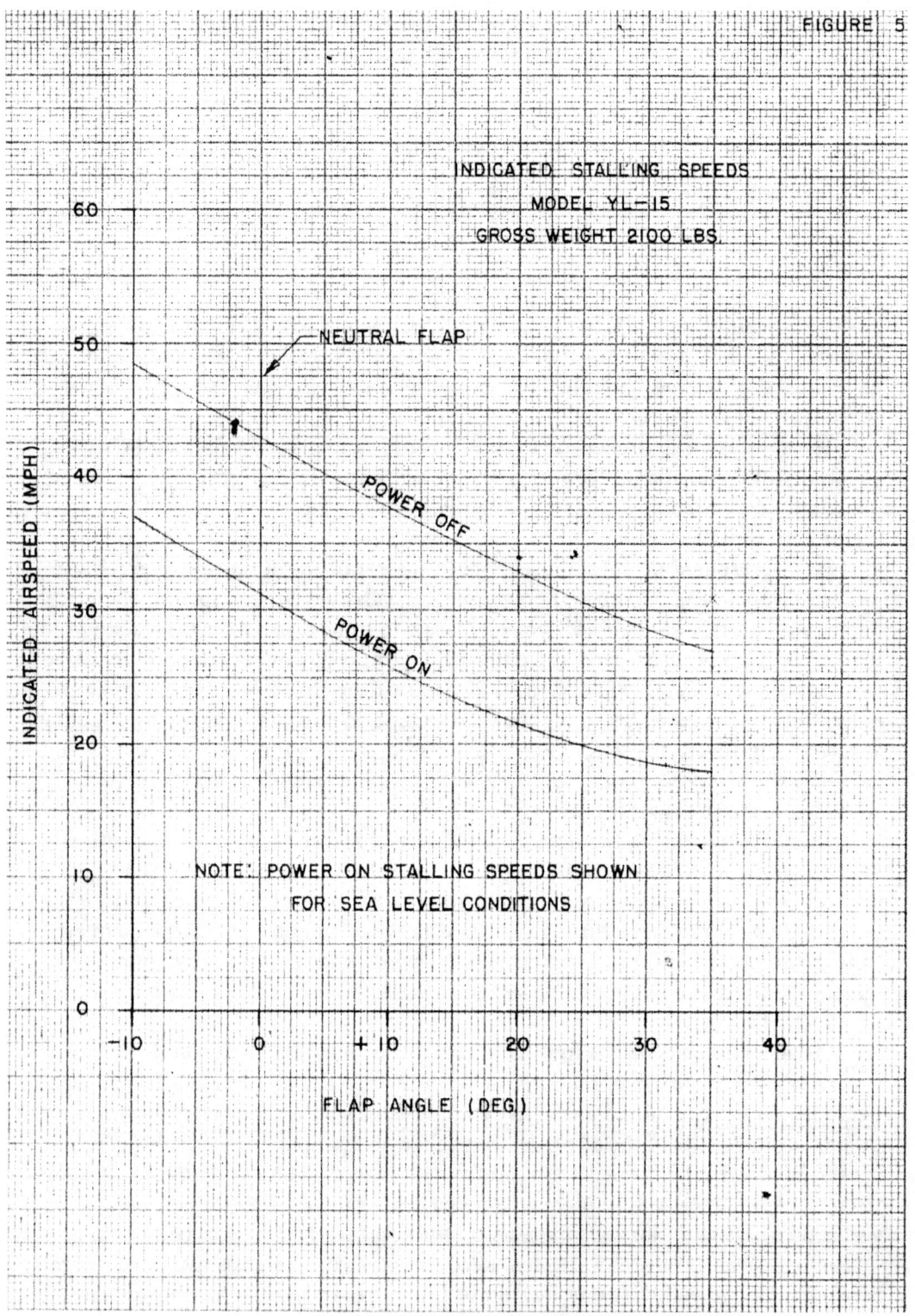

YL-15 stall speeds (author's collection).

non-recovery or produce a dangerous spin, but nothing tried would even appreciably slow down the positive recovery characteristics. With 20° flaps or more, a spin entry could be made, but the aircraft would then go into a spiral at about the one-half turn point and could be flown out of this spiral at any time.

The elevator was designed to be extremely effective when operating at very low airspeeds, which resulted in it being quite sensitive at high airspeeds. Likewise, the large elevator trim tab also becomes sensitive at high airspeeds. Prior to any maneuver involving a diving speed of 125 mph IAS or more, it was recommended that the trim tab be placed at a position slightly "nose heavy" from the level flight setting to avoid light stick forces and the potential of applying high acceleration or "g" loads during the pull-out. The V-G diagram should be consulted for the allowable "g" loads that could safely be developed at various airspeeds. One of the specifications for the airplane was that it be stressed for a maximum dive speed of 200 mph, or approximately 100 percent above the level flight top speed, whereas most conventional airplanes are stressed to a maximum dive speed of roughly 25 percent to 50 percent of their level flight top speed.

On the ground, the pilot might notice a spring force resisting the sideward motion of the stick. Since the lateral control spoilers gave no air load on the stick when deflected, the spring force was included to simulate normal air loads at approximately cruising speed. At low flying speeds, the spring force gave a rather unnatural aileron "feel" in that stick forces did not reduce appreciably with speed; however, it was thought that this would not prove to be objectionable as a pilot's experience in the airplane built up.[6]

Landing

The YL-15 had excellent elevator control throughout the low-speed range, providing good landing characteristics. The spring-oil shock absorbers on the landing gear absorbed a great deal of the initial landing shock, resulting in readily attainable smooth landings.

Approaches and landings with 0° flaps were normal and compared favorably with other airplanes of this general weight-horsepower class. With 20° or more flaps, the attitude of the airplane on the approach glide became extremely steep and required a large angular change from the approach to the three-point landing attitude.

Full stop landings over a 50-foot obstacle had been made in 330 feet (as corrected to standard day, no wind, and sea-level conditions). These were accomplished with full flaps, practically stalling the airplane just over the obstacle out of a straight glide, then diving it steeply to gain enough speed to accomplish the flare-out for landing, applying power as necessary to make the flare-out safely, and using full brakes just after ground contact was made. This was the procedure also given in the flight handbook.

Slipping, fishtailing, or other maneuvering on the approach was not necessary to make consistently short landings as the airplane was designed and equipped with large flaps to accomplish short landings easily. Boeing believed that by using care in selecting the proper flap setting for the flight condition at hand (wind, obstacles, field length, terrain, etc.) and by controlling both the airspeed and the nose attitude closely in a relatively steady wings-level glide approach, short landings could be made consistently with less effort than if by using slips or other maneuvers.

During a landing approach from which a go-around might be necessary, the airspeed should be kept high enough so that a good rate of climb could readily be attained upon power application. The minimum indicated airspeeds were roughly 50 mph at 0° flaps and 45 mph at full flaps to maintain this condition. Steeper approaches could be made by maintaining slower airspeeds, but unless the speeds were kept higher than the above minimums, or at least allowed to build up to them when a go-around appeared to be necessary, there would be very little rate of climb possible even with full power. For an extended climb-out after the initial go-around had been completed, the flaps should be raised slowly to a maximum setting of 20° or less when a safe altitude had been reached. This was because the L-15 had a power-required curve shaped somewhat differently from that of other light aircraft.

Full brakes could be applied as soon as ground contact was made on a landing with little or no tendency to nose over. This was because almost as much of the airplane weight was supported by the tail wheel as by each main wheel.

The pilot needed to remember landings could be made in roughly one-half the distance required for takeoff; hence, selection of landing areas should be normally made on the basis of the takeoff distance rather than the landing distance.

High gusty winds often had the effect of tending to roll or yaw the airplane on its landing run when flaps were used. This effect was largely due to the extremely low flap deflected stall speeds which were below gust velocities often encountered in winds slightly above average. A difference of just a few miles an hour between the gust velocity on the two wings could easily result in one wing being in a stalled condition while the other still had considerable lift. The airplane would exhibit a "wallowing" tendency under these conditions. Apparently, the best remedy was immediately raising the flaps to 0° on contact with the ground, then to the negative 10° setting as soon as possible. After this was done, the airplane normally settled firmly on the ground and could readily be controlled by the steerable tail wheel. Due to the tail surfaces being located out of the propeller slipstream, using the throttle to gain directional control was ineffective; thus, firm ground contact with all three wheels was desirable as soon as possible.

Landing characteristics with 0° flaps in high gusty winds compared favorably with those of other airplanes in this general weight-horsepower class. Hence, judicious use of flaps in proportion to the wind and gust velocity should result in above-average landing characteristics and handling qualities. Since the landing distance required by the L-15 was unusually short with 20° or more flaps and no wind, the distance needed with 0° flaps in a high wind should not be much greater; hence, little sacrifice in landing distance should be necessary by suiting the flap setting to the weather conditions.[7]

Power Required Curve

The power required curve shows the engine power (given in terms of rpm) necessary to maintain a given true airspeed in level flight at standard day condition.

One of the most important things about a power required curve from the pilot's standpoint is its shape. The one on the L-15 was extremely steep when compared to those of other light aircraft.

The shape of the XL-15's power required curve from 53 mph to high speed was different from those of most other light airplanes in that it tended to "hook" outward at 88 mph. This hook was due to the thin metal blades of the McCauley propeller tending to bend and increase their pitch slightly at high engine powers and airplane speeds, giving an effect similar to a constant-speed propeller. A slightly higher speed could be attained for a given rpm at this region of the curve. The curve below 53 mph is exceptionally steep and rather unusual. This low-speed portion of the curve is often referred to as the "backside"—an airplane flying at these conditions is "hanging on its prop" with considerable power at very low airspeeds. Since this portion of the curve on most airplanes is comparatively shallow, such airplanes can be slowed down appreciably (often well below their "power off" stall speeds) by using considerable power at very low speeds—this condition can be used to advantage in maximum performance, short field landing approaches. However, on the L-15 it can be seen that addition of power below about 47 mph does not result in decreasing the airspeed appreciably; also, it was found in flight tests that the use of power on a landing approach, particularly with flaps at 20° or more, decreased the rate of descent—thus making the angle of approach shallower. This indicated that the steepest landing approach could be made with "power off" and full flaps.

The lowest point on the curve (approximately 53 mph) was the speed at which level flight could be maintained with minimum power. It was also the dividing point between a condition where more power would give more speed or one in which more power would result in less speed—the speed being determined by the nose attitude. This had to be considered on landing approaches from which a go-around might be necessary. If the airplane was brought in on the approach with 0° flaps at much less than 53 mph (approximately 48 indicated airspeed) and a go-around was necessary, the addition of full power without an increase in true speed to at least 50 to 53 mph would result in a condition where most of the power was necessary just to maintain level flight and, hence, almost no climb would be possible. This same condition would exist with flaps deflected—but at correspondingly lower airspeeds.

Power required curves slightly change with altitude—at a higher altitude, the curve would be higher than, and inside the one shown, but would tend to cross it at a slight angle at the high-speed end; however, the general shape and characteristics are similar.[8]

Cruise Control

From the pilot's standpoint, cruise control with a fixed-pitch propeller consists of proper and precise fuel mixture leaning and maintaining constant indicated

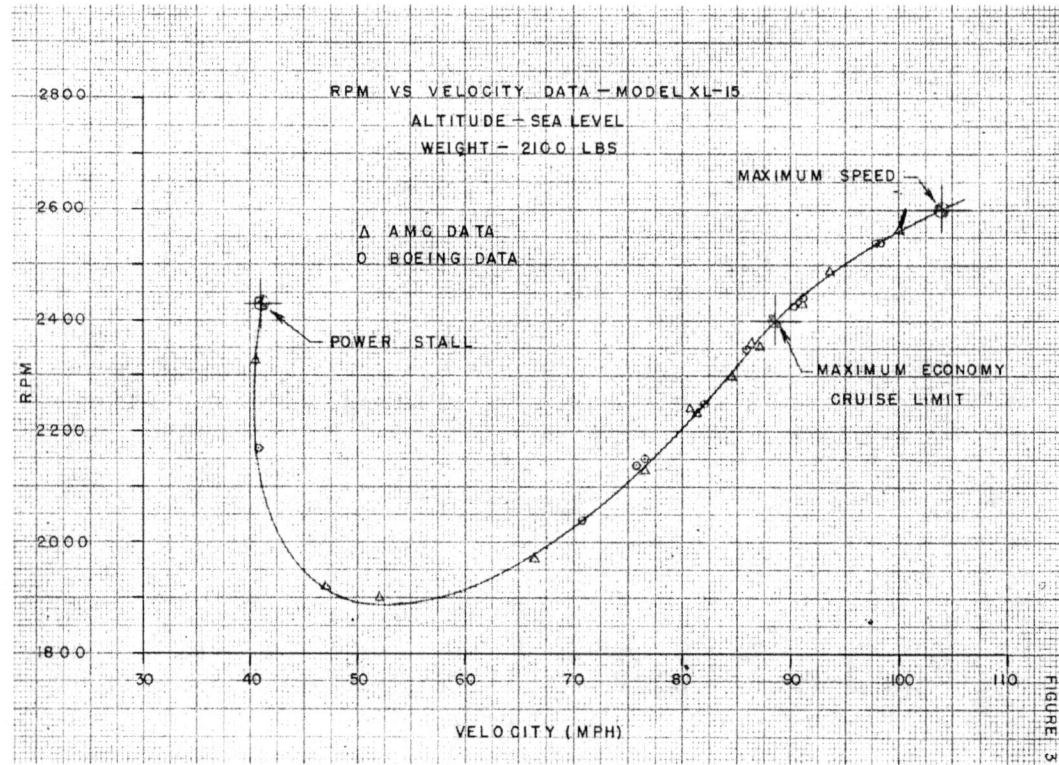

XL-15 power required curve (author's collection).

airspeeds. This could be used to advantage to conserve fuel and gain additional range or endurance. Doug Heimburger stated that the proper control of fuel mixture and airspeed would increase range and/or endurance by an *amazing amount*. With the standard wing tanks filled, an absolute range of 335 miles or an endurance of 5.2 hours was possible, whereas, with the external tank, an absolute range of 760 miles or an endurance of 12.7 hours could be obtained. These figures apply to standard day, zero wind conditions. These data were taken from Air Force and Boeing Wichita flight test data and had been reduced by 5 percent to allow for practical operating conditions.

The Lycoming 0–290–7 engine installed in the YL-15 had no altitude or power restrictions in leaning the mixture to "best power" on; however, leaning to "maximum economy" was recommended only at less than 65 percent horsepower or at indicated airspeeds of less than 87.5 mph with a 0° flap setting. The "best power" mixture setting produced maximum engine power by providing the best power fuel-air mixture for the given atmospheric condition. The "maximum economy" mixture setting provided power with minimum fuel consumption by supplying a lower fuel-air ratio to the engine.

To lean the mixture to "best power" for any given throttle setting, the indicated airspeed needed to be held constant to maintain a constant engine rpm, then the mixture control lever slowly moved aft until the maximum rpm rise had been obtained. The mixture control lever would usually have to be moved slowly back and forth several times around the position that gave maximum rpm until the exact setting was determined, and normally a slightly "rich" setting was favored. It was suggested that this procedure be used on any flight in which a considerable quantity of fuel was to be used. To lean the mixture to "maximum economy," the "best power" mixture setting should be obtained, and then the mixture control lever should be moved slowly further aft until a drop of 60 rpm is obtained while the indicated airspeed is held constant. Over-leaning the engine should be avoided.

Altitude affected the mixture setting, which must be readjusted for altitude changes; however, once properly adjusted at any one altitude, the mixture setting would hold good for any throttle setting. Except from fields at very high elevations, takeoffs and landings should be made with the mixture at "best power."

When the engine rpm was reduced to obtain a "maximum economy" mixture setting, the airspeed necessary to maintain level flight would also be reduced. Desired indicated airspeeds, of course, could then be maintained by normal adjustment of the throttle to a constant position.[9]

Aircraft Load Limits

Not knowing the experience level of the pilots that would be transitioning to the YL-15, Doug Heimburger included a section on the aircraft's load limit. The plot of allowable velocity vs. "g" load (true indicated airspeed versus acceleration or "g" load) was included to show the structural limitations of the airplane. To find the allowable "g" load for a given true indicated airspeed, the pilot read vertically from the speed to the positive or negative curve, then horizontally to the "g" scale. (Note that today, true indicated airspeed is called calibrated airspeed.)

The curve from the level flight stall speed at one "g" to 86 mph indicated the speeds at which the airplane would stall under the corresponding "g" load; hence, it could not readily be exceeded. From 86 to 112 mph, the limit on the upper or positive "g" curve was plus 4 "g's"; on the negative or lower curve, the limit was minus 2 "g" between 98 and 112 mph; above 112 mph, the allowable "g" load decreased with increasing airspeed as shown until the limit of 200 mph and plus 2.47 "g" was reached. All flight maneuvers were to be made within this "V-G envelope" to avoid breaking or at least bending the structure. A factor of safety of 1.5 above the "limit load" shown on the diagram was built into the airplane—the structure could reach its ultimate load and break if this condition was reached or exceeded. The dashed straight line starting at one "g" and 0 mph indicates the "g" load that could be imposed at various speeds by a standard maximum atmospheric gust. Static tests at Wright Field proved the structure of the L-15 to be extremely efficient.

The structural integrity tests flown as part of the Phase III tests on the second XL-15 proved that the airplane would meet the limit loads of the V-G diagram shown without affecting the structure. In addition, a condition was recorded which proved the ability of the L-15 to withstand a 4 "g" load at its maximum flight towing speed of 165 mph.

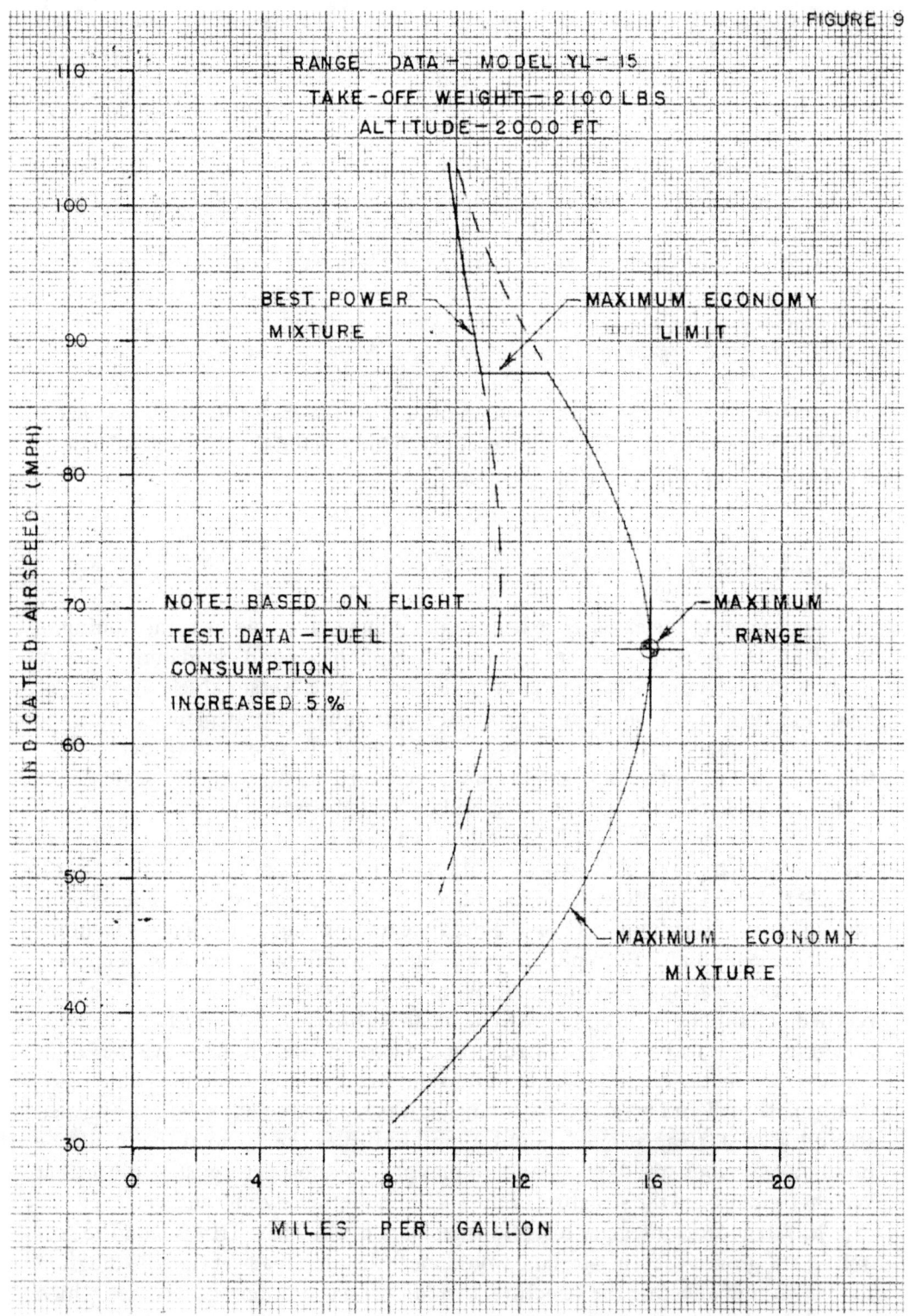

YL-15 range data (author's collection).

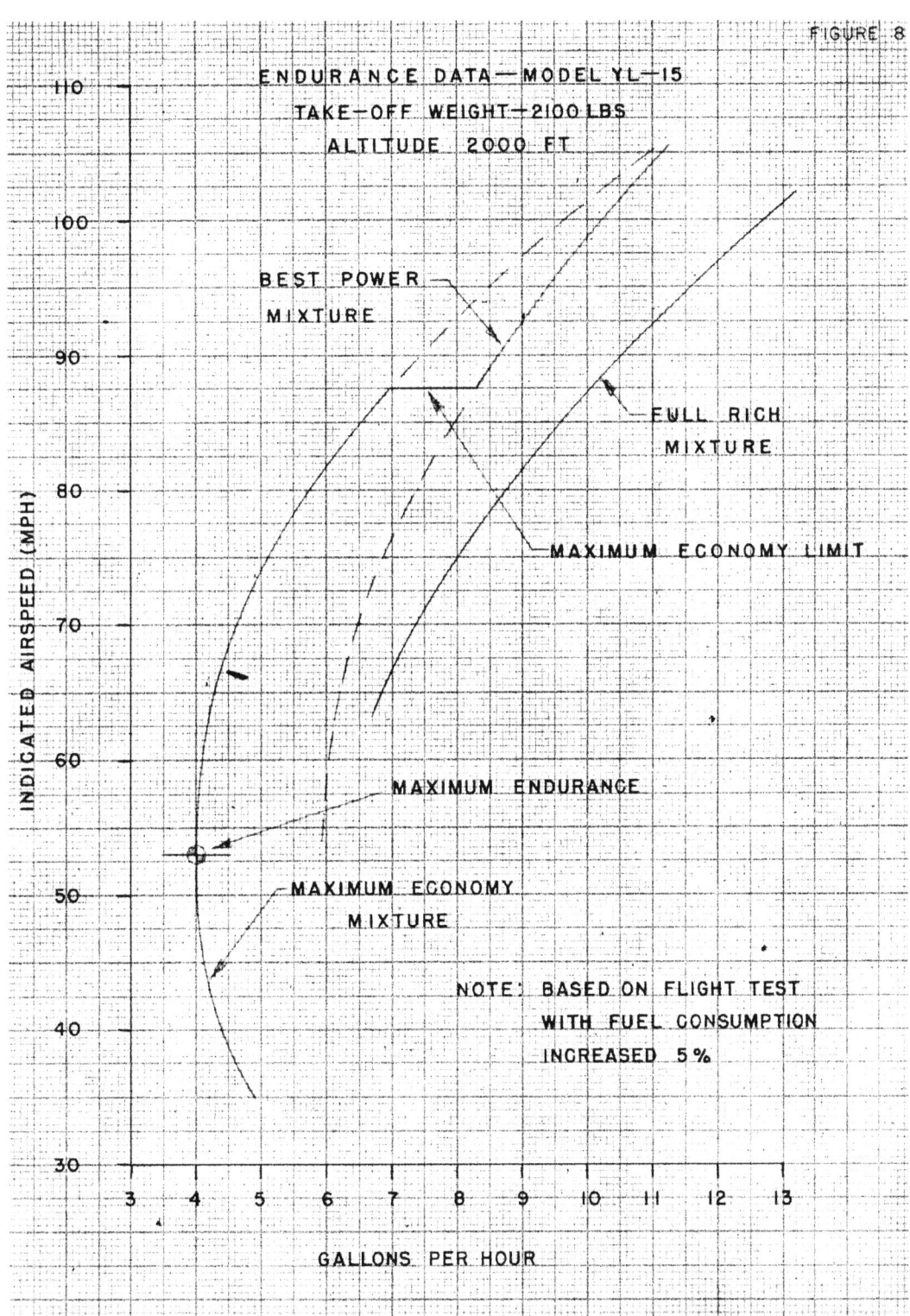

YL-15 endurance data (author's collection).

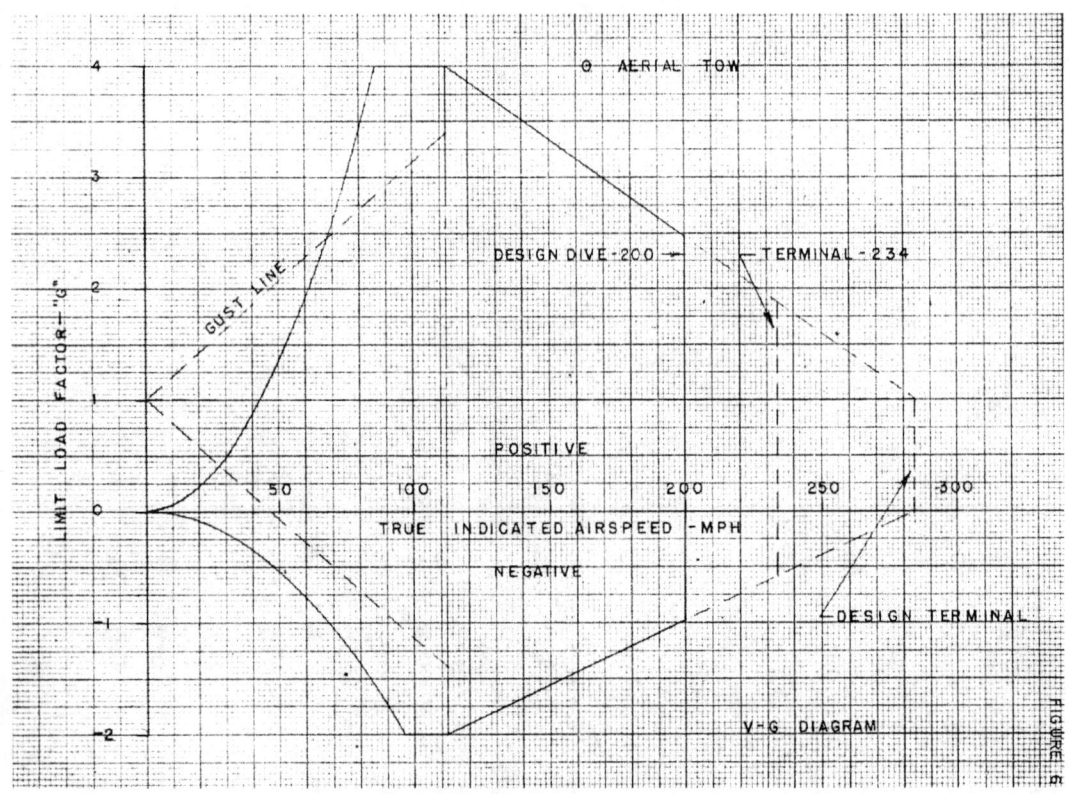

V-G diagram (author's collection).

The terminal velocity of the YL-15 had been computed to be 232 mph true indicated airspeed from the results of the dive test to 200 mph that was flown on the second XL-15. A hypothetical terminal velocity of 284 was used for structural analysis. At gross weights above standard the "g" load limits will be lower.[10]

14

USAF YL-15 Accelerated Service Test

A service test evaluates any problems in the operational and tactical use of an aircraft and is usually done with a number of aircraft under field conditions. The U.S. Army did the service test with nine YL-15s at Fort Bragg.

However, it is normal to give the first production item a much shorter accelerated service test of test conditions designed to reproduce in a short time the deteriorating effect obtained under normal service conditions. This is done with experienced crews in relays that give an airplane 150 hours of almost continuous flight with a full military load. In a very short time period, this accelerated testing brings to light mechanical defects that normally might require years of service experience for discovery. Even though the L-15 was to be an Army aircraft, it was developed under the specifications and guidance of the Air Force. Thus the U.S. Air Force would do the accelerated service test, using the first YL-15 47–423.

The test was documented in Air Materiel Command Memorandum Report MCRFT-2213, "Accelerated Service Test of YL-15 Aircraft," prepared by Steven G. Steffens, Engineer, Accelerated Service Test Branch, Test Engineering Subdivision, dated April 12, 1949. The accelerated service tests on the aircraft began on December 3, 1948, and were completed on April 6, 1949. A total of 150:15 flight test hours were accumulated during this period. The initial part of the test was done in Wichita by Air Force personnel at the Boeing plant so they would have the Boeing technical personnel at hand to draw knowledge from if needed. On December 22, 1948, aircraft 47–423 was flown by a USAF pilot from Boeing Wichita to Wright Field for the main part of the accelerated service testing.[1]

Factual Data

Many of the items identified in the report were just things that needed a slight adjustment, and these were noted, and the adjustment described. This is sort of like the nitpicking you do when you get a new car. Other items detailed were parts that broke or wore out and needed repair, replacement, or even a small modification or redesign. Lastly, those items that could be lived with on this first batch of aircraft but should be changed in any future production aircraft were noted. Also, man-hour records were kept to determine the workload required for various types of inspections.

YL-15 47-423 was the accelerated service test aircraft (NARA via AEHS).

The report was very detailed in its factual data section, and only the more troublesome or significant findings will be discussed here.

Inspection of the elevator tab, P/N W2-320003, at 25:00 test hours showed the tab had warped 4°. This warpage was due to the light construction of the tab and to its length.[2]

During an assembly of the aircraft, after a tear down for towing purposes, it was found impossible to connect the flap and aileron cables to the quick-disconnect, P/N RA 2258-3, without loosening the cable turnbuckle. It was difficult to unfasten the cables from the quick disconnect when rigging the cable tensions according to the applicable technical orders.[3]

Numerous maintenance difficulties were experienced with the pilot's entrance door.

    (a) The sliding Plexiglas window in the door cracked several times during flight because, when partially opened, the panel fluttered and eventually cracked. After several failures of this type, the sliding panel was replaced by a fixed panel, and no further failure occurred.

    (b) The door was hinged on the backside so that it opened into the propeller blast. Several door windows were cracked by pilots opening the door while taxiing and having the door pulled from their grasp by the propeller blast.

    (c) During one flight, the pilot's entrance door was blown off because the door latch had not locked in place. The spring, P/N W2-410867, which locked the latch was too weak to provide a positive-looking action. A modification provided by Boeing, which incorporated a heavier spring, corrected this difficulty.[4]

The forward view of the observer was obstructed by control cable shield, P/N W2–416377. Also, with this shield in place, it was extremely difficult for the observer to fly the aircraft with the observer's flight controls. A shorter shield designed to protect the bell crank, pulley, and cable assemblies located on the floor between the pilot and observer would provide the necessary protection from fouling the controls and would provide the observer with additional forward vision.[5]

The accessibility of the various components of the aircraft was, in general, excellent. However, an improvement in the accessibility of the various flight control bell crank and pulley assemblies located in the tail boom between the fifth and sixth bulkheads, counting from the base of the boom assembly, should be made by installing a stressed inspection plate in that area to facilitate inspection and maintenance.[6]

The ground handling of the aircraft was satisfactory. It was noted that three men could easily push the airplane and control its direction.[7] (Author's note: I wonder what the Army thought about this, as the L-4 it was to replace could be moved about by one man on the ground.)

The intervals and approximate number of man-hours required to conduct the various aircraft inspections was reported as shown in Tables 14.1 and 14.2.[8]

Table 14.1

Inspection Interval	*Hours*
25-hour inspection	10 man-hours
50-hour inspection	25 man-hours
100-hour inspection	30 man-hours
Daily and preflight	1 man-hour

On the missions flown, the aircraft was at takeoff gross weights of from 1,920 pounds to 2,119 pounds and the CG location from 36.5 percent to 42.3 percent MAC. The types of missions, hours flown, and type and number of landings were as shown in Tables 14.2 and 14.3.[9]

Table 14.2

Type of Mission	*Hours*
Familiarization	48:35
Long Range Ferry 2,000 feet Altitude	15:16
Long Range Ferry 6,000 feet Altitude	10:15
Minimum Air Speed Missions	15:40
Maximum Continuous Power 5,000 feet Altitude	12:50
Maximum Continuous Power 8,000 feet Altitude	12:00
Altitude Mission 10,000 feet	4:10
Service Ceiling	6:00
Evasive Tactics	9:30
Night Missions	9:00
Detached Service Flights	7:00
Total	150:15

Table 14.3

Type of Landing	Number
Sod	384
Snow	58
Concrete	13
Total	455

The Air Force pilots expressed satisfaction with the general layout of the cockpit. They stated a step should be installed on the right landing gear strut to eliminate the possibility of the pilot's foot slipping on the sloping strut. The seats were noted as comfortable, and the instrument location and grouping as satisfactory. The rudder pedal adjustment was adequate, and the stick position was satisfactory. Doubt was expressed as to whether the two ventilation ports would be adequate to keep the cabin cool during the summer or in tropical areas. The heater was deemed marginal in temperatures of –10°C or less. The radio setup was excellent, and the location of the flap handle, parking brake, throttle, and electrical switches satisfactory.[10]

Aircraft control during crosswind taxiing in winds under 20 mph was satisfactory, with the use of negative flaps aiding in the ground control of the aircraft. However, in winds above 20 mph, excessive brake use was needed to control the aircraft during crosswind taxiing.[11]

The Air Force pilots had several observations on handling characteristics in flight:

(a) At slow airspeeds of 40 to 45 mph, response to aileron was slow, and the aircraft tended to wallow laterally.

(b) In level cruise flight, all controls were adequate; however, at altitudes of over 8,000 feet, the response to controls was slow.

(c) The stall characteristics were normal. The aircraft stalled straight ahead, giving adequate warning and easy recovery.

(d) All pilots considered the crosswind landing characteristics of this aircraft to be unsatisfactory. Rudder and aileron control at landing speeds was inadequate. Response to aileron control was very slow, and loss of rudder control on landing occurred too soon. Use of opposite brake was needed to hold the aircraft rolling straight. Cross-wind landings in steady winds up to 20 mph were considered safe. Above wind velocities of 20 mph, the margin of safety dropped rapidly. In gusty wind conditions, landings were definitely hazardous.

(e) Rudder control during crosswind takeoffs was definitely improved by not lowering the flaps until 30 to 35 mph airspeed was obtained.[12]

The pilots observed that during night flying, the intensity and the angular adjustment of the landing lights for takeoffs, landings, and taxiing were satisfactory. Instrument lighting was satisfactory, but some pilots suggested more illumination of the instruments in the center of the panel and the standby compass. A particular concern of all the pilots was the loss of forward vision during night landings in rain. The rain adhered to the Plexiglas windshield, thus obstructing and distorting the pilot's forward vision. All other items were satisfactory.[13]

Entrance into the observer's station was easy. The seat was comfortable and all-around visibility excellent except for forward visibility, which was blocked by the cable guard installation between the pilot and observer. Flying the aircraft from the observer's position was possible but not practicable, owing to the restricted forward vision of the observer, even with the cable guard removed. A footrest installation, when the observer faced to the rear, would increase his comfort.[14]

Conclusions

The memorandum report stated that, based on the accelerated service test, the following conclusions were made:

(a) The YL-15 was satisfactory as to maintenance, since no major maintenance problems were encountered during the test, and the mechanics found the accessibility features of the various components excellent. Many of the minor maintenance problems could be eliminated by improved contractor inspections.

(b) The utility of the aircraft for liaison purposes was limited because of several unsatisfactory operational characteristics, the most important of these being the crosswind landing limitations. Landings in crosswinds of 20 mph or greater were found hazardous, and gusty wind conditions further reduced the maximum wind velocities allowable for landing. Inadequate aileron control at slow speeds and the early loss of rudder control on landing contributed to the operational shortcomings of the aircraft. The critical icing characteristic of the carburetor was also a hazardous condition.[15]

The accelerated service test consisted of 150 flight hours done in four months (NARA via AEHS).

YL 47–423 was a U.S. Army airplane, but the accelerated service test was performed by the U.S. Air Force (NARA via AEHS).

Recommendations

On the basis of the results obtained during the service test, the Air Force recommended,

    (a) The contractor improves the assembly and inspection of the following units which were found improperly assembled on the test aircraft.
        (1) The fuel gage float.
        (2) The flap control bell crank assembly.
        (3) The rudder control bell crank and cable assemblies installed inside of the vertical stabilizers.
    (b) The fuel screen one-fourth inch pipe housing, which screws into the carburetor float chamber, be made longer so that the elbow on the end of the short pipe housing will extend beyond the lower engine crankcase. This would make the removal and inspection of the finger screen easier.
    (c) The external power receptacles be made with insulating strips of phenolic material.
    (d) The sharp points on the two travel wedges of the mast assembly, tail wheel steering, be rounded to prevent cutting into the spindle housing plate.
    (e) The elevator tab be more rigidly constructed so as to prevent the warpage experienced on this unit during the test.
    (f) The aileron-feel clamp and aileron-feel eye-bolt be redesigned so as to

eliminate the aileron-feel spring hooks chafing in the aileron-feel clamp and aileron-feel eye-bolt eyelets.

(g) An enclosure be built around the various control assemblies located on the cockpit floor between the pilot and observer, to prevent jamming of these controls by objects which may roll back underneath the pilot's seat and into the controls.

(h) The quick-disconnect, P/N RA2268-3, be redesigned or a special tool be developed that will pull the control cables together to aid in inserting the swaged ball end of the control cables into the quick-disconnect.

(i) The pilot's entrance door be hinged at the front end to eliminate the potential danger of the door being blown open and off during flight or during taxiing.

(j) A water drain hole be drilled into the lower section of the nose-assembly-engine-cowl.

(k) The water tightness of the observer's cockpit be improved where the tail boom is joined to the fuselage and around the Plexiglas sections of the cockpit roof.

(l) Greater care be taken when installing the various engine air baffles to prevent the chafing end cracking of these units.

(m) The bracket securing the center aft-canopy-retainer strip to the top fuselage longeron, above the observer's compartment, be made of heavier gage material.

(n) Hexagonal nuts, instead of knurl nuts, be used to secure the high tension shielding around the high tension leads.

(o) The anchor bolt, which secures the safety belt to the pilot's seat, be modified by cutting threads on both ends. Removal of the safety belt would then be simplified because only the exposed anchor-bolt nut on the side of the pilot's seat would need to be removed.

(p) The control cable shield, P/N W2-416377, be eliminated or the length reduced, to improve the observer's forward vision.

(q) An inspection plate be installed in the aft section of the tail boom to permit inspection of the various flight control bell cranks, cables, and pulley assemblies.

(r) The inspection of the air valve assembly be improved to eliminate the discrepancies found on the test aircraft installation, and that the air valve shaft, P/N W2-640048, be fabricated from steel.

(s) The fuel mixture-control linkage be redesigned to obtain adequate clearance between the control-link rod and carburetor housing.

(t) A step be incorporated on the landing gear strut to provide a safer foothold for the pilot when entering the pilot's cockpit.

(u) Additional air vents for cabin cooling be provided, as the present installation is marginal.

(v) A study be made of the possibilities of using a tricycle landing gear configuration to improve the unfavorable crosswind takeoff, taxiing, and landing characteristics of this aircraft.

(w) The ailerons and rudders be redesigned to provide better control at slow speeds.

(x) Modifications be made on the front windshield to eliminate the serious loss of forward vision due to rain. The problem is quite serious when landing.

(y) Footrests be installed for the comfort of the observer when facing to the rear.

(z) An improved type of fuel metering device be installed to reduce the serious carburetor icing encountered on the present installation.

Consideration be given to engine installations of greater horsepower to improve the climb and altitude performance of this aircraft.[16]

15

The Army, the Air Force, and the L-15

World War II Intraservice Battle over Liaison Aircraft

The Army Ground Forces and the Army Air Forces (later U.S. Air Force) had a controversial relationship over liaison aircraft. It began just before World War II when the Army Ground Forces came to the conclusion they needed aerial support in front-line observation, including aerial direction of artillery. The Army Air Forces were of the opinion they should be operating and controlling all aircraft. At the time, the Army Ground Forces were willing for the Army Air Forces to do those tasks;

Piper J3C loaned to Army in 1941 maneuvers (author's collection).

however, the Army Air Forces repeatedly failed to do them. For the 1941 maneuvers, the Army Ground Forces, in addition to asking the Army Air Forces for observation aviation support, borrowed airplanes and pilots from three light plane manufacturers. The Army Air Forces support was a big failure, while the loaned light planes' mission performance turned out much better than had been expected.

The Army Ground Forces using the maneuver results convinced the Secretary of War to allow them to buy light planes (primarily the Piper L-4) and attach them directly to artillery units. The actual aircraft purchase was made through the AAF's Air Materiel Command. However, the Army Air Forces would not give up the fight for control and formed Liaison Squadrons and developed an aircraft, the Stinson L-5, to equip them.

The L-4 was a 65 horsepower Piper Cub with enlarged windows and had spectacularly short takeoff and landing distances. The L-5 was larger, with three times the power giving it much better climb, speed, and altitude performance, but its field performance was not as good as the L-4.

Although both the Army Ground Forces and Army Air Forces units were liaison aircraft operations, they were operated and controlled in two different ways.

The Army Ground Forces aircraft were operated from small strips generally adjacent to artillery batteries and under the control of the local ground organization. The aircraft were used for targeting artillery, checking camouflage, convoy direction, and personnel transportation. Army Ground Forces liaison pilots were generally artillery officers that were trained as both pilots and mechanics. In addition, mechanics to maintain the aircraft were also assigned to the artillery air section.

Piper L-4 operating off a road during the Korean War (U.S. Army).

Stinson L-5 (AAHS).

The Army Air Forces liaison squadrons were operated primarily from air bases and under the direct control of the Air Force. The aircraft were used mainly for communications, mail, and staff transportation. The Army Air Forces liaison pilots were sergeants and trained only as pilots. Aircraft maintenance was generally done at air depots. The Army Air Forces liaison squadrons would do the same missions as the artillery section liaisons, but only when requested by a ground commander through the Air Force command organization, and the Army Ground Forces had to provide an air observer to fly with the Army Air Forces liaison pilot. Often this meant an Army Ground Forces requested mission would not get flown until after the need or battle was over.

This dual set of liaison aircraft organizations existed throughout World War II, and for the entire time, the Army Air Forces did everything they could to get Army Ground Forces aviation shut down.

After the invasion of Italy, when the U.S. Army was fighting in the Italian mountains, it became very obvious that the L-4 was underpowered for operation from higher terrain. After some bitter discussion, the Army Air Forces agreed to allow the Army Ground Forces to be issued a few higher-powered Stinson L-5s.[1]

New Liaison Aircraft

About 1943, the Army Ground Forces seeing the need for improved liaison aircraft to replace the Piper L-4, and, as would be expected, the Air Materiel Command at Wright Field showing little interest in supporting the Army

Piper L4X, prototype of the L-14 (NARA).

Ground Forces' desire, the Army started to look on their own at what might be available. Without Air Force involvement, they could not officially request a new aircraft to be developed, but they could evaluate experimental aircraft that light aircraft manufacturers had developed at their own expense. The artillery center at Fort Sill went so far as to organize several flight evaluation days where prototype liaison aircraft were examined. Out of this type of activity, the Piper L-4X was determined to be the best candidate. Against the Army Air Forces' objections, the Army Ground Forces got the Secretary of War to authorize the purchase of 850 of the Piper L-4X, designated as the L-14, to be the new standard field artillery aircraft. The L-14 was powered by the Lycoming O-290, which would later be used by Boeing in the L-15.[2]

By the time the war with Japan ended, five YL-14s had been completed, and the remaining 845 were canceled. But because of a new understanding between the Army Ground Forces and the Army Air Forces as described in the following paragraph, a new purpose-designed liaison aircraft (L-15) would be procured for the Army in place of the canceled L-14.[3]

Postwar

As the war ended, it was getting obvious that the Army Air Forces was gaining political backing to have itself separated from the Army and become a separate armed service. But to achieve this without a vicious battle in Congress, they would need to have the blessing of the Army Chief of Staff. It was pretty certain that General of the Army Dwight D. Eisenhower would become the next Army Chief of Staff. Ike had been one of Army aviation's biggest backers and let the Air Force staff know that he would not block the separation, providing that the Air Force did not push for the doing away of organic Army aviation.

The Army Air Forces immediately reversed tactics from trying to eliminate aviation in the Army Ground Forces to becoming somewhat supportive. Since the Army had no aircraft development engineering organization, the Air Force would engineer and develop any new aircraft for the Army. The Army was allowed, however, to make purchases on their own of off-the-shelf commercial light planes that did not require development.

The Air Force, after separation, did, however, continue having their own liaison squadrons for many years. These, however, became basically light transport and communications services for internal USAF use.

Air Force Liaison Aircraft Competition for the Army

Just months after the end of World War II, the AAF (soon to be the USAF) Air Materiel Command began working on the informal liaison aircraft competition that would result in the L-15. The Request for Bids that the Air Materiel Command sent out based on the Army's "Military Characteristics of Field Artillery Observation Aircraft, dated 20 August 1945," and the technical requirements included would surely have been ones the Army would be happy with.

It is in who would judge the bids that you can see an Air Force view that "we know airplanes and the Army does not." In the performance section, the points were based on equations that mathematically rated the bid performance against the desired and minimum acceptable performance and thus gave a fair rating of performance. In the engineering sections (structures and design, power plant installation, maintenance and repair, and equipment installation) all the judges were Air Force engineers. Only in the suitability (operational utility) section did the Army get a vote, and it was a minority vote as the points were determined by a committee consisting of Wright Field Air Force engineers and one Army officer, the Army Ground Forces Liaison Officer assigned to the Air Materiel Command.

One would have thought that since this aircraft was being developed for the Army, they would have had more say in the selection of the winner. Despite the Air Force viewpoint that the Army "did not know airplanes," many Army Ground Forces aviators were degreed engineers, and at that time, all were trained as pilots and in the repair and maintenance of light aircraft.

Mockup Inspection

The mockup inspection was held on 3–6 September 1946 at the Boeing Wichita facility. In addition to many Wright Field civilian engineers, the following officers were present:

(a) Army Air Forces (all members of the Mockup Board)
Lt. Col. J. A. Hogg

Lt. Col. J. G. Earnest
Major D. E. Riley
Major T. M. Ruckman

  (b) Army Ground Forces
Col. B. Evans, AGF Development Section—Advisory member of Mockup Board
Col. W. W. Ford—Advisory member of Mockup Board
Lt. Col. C. W. LeFever, AGF Liaison Wright Field—Member of Mockup Board
Lt. Col. C. L. Sheppard–Observer
Lt. Col. Jack Marinelli–Observer
Major D. L. Bristol—Observer

The Army Ground Forces did get a larger presence on this inspection, even though only one was a full member of the mockup board.[4]

Checking observer's accommodation during Mockup Inspection (NARA).

Army Lets Its Troops Know about the L-15

In *The Field Artillery Journal* of October 1946, the XL-15 was announced with drawings (see images on pages 55 and 59), a three-view, and estimated performance. The journal stated that the initial models would be subjected to extensive tests before quantity production started. It said there were many reasons to believe that in the not-too-distant future, the Ground Forces would have a light aircraft far superior to any used in the past.[5]

The Army also let the general public know about the L-15 by featuring it in popular magazine advertisements.

Second Thoughts on the L-15

In December 1946, before the first XL-15 was even built, Major General Charles L. Bolte, the Army Ground Forces Chief of Staff, sent a letter to the Commanding General, Army Air Forces, stating that the existing Army Ground Forces liaison airplanes were getting old, with many approaching the end of their useful service, and that the XL-15 which was originally scheduled to replace these aircraft was still in the early stages of development and would not be capable of starting quantity production until at least December

New "eyes" for the Army Ground Forces

You might guess this little aircraft belonged to the Air Forces, but you'd be wrong. It's strictly a Ground Forces gadget — the latest development in liaison and spotting planes.

It's the 125-hp. L-15, built to do the almost unbelievable things required by the Infantry, Artillery and Armored Cavalry.

This plane can cruise at a top speed of 112 mph., and dive safely at 200, yet it can hover at 36 mph. without stalling! On a field where there are no obstacles it can take off or land in less than 250 feet — or take off and clear 50-foot trees in 600 feet.

One of the L-15's most outstanding features is full visibility in all directions. Pilot and observer have a sweeping, all-round view, including straight down and straight aft, made possible by the unique cockpit design and elevated tail spar.

The role of small observation planes in World War II was vitally important to victory. Spotting enemy installations, they enabled our firepower to strike with deadly accuracy. This new plane will give the Ground Forces still more dependable "eyes."

All through the peacetime Regular Army you'll find new scientific advances being tested and perfected. It's a great career for young men of mechanical skill who want to get ahead. There's good pay while learning; food, clothing, shelter and medical care; travel and splendid educational opportunities. A 3-year enlistment permits you to choose your branch of service from those still open. Ask for full details at your nearest U. S. Army Recruiting Station.

● Listen to "Sound Off," "Voice of the Army," "Proudly We Hail," "Front and Center," and major football broadcasts on your radio.

A GOOD JOB FOR YOU
U. S. Army
CHOOSE THIS FINE PROFESSION NOW!

YOUR REGULAR ARMY SERVES THE NATION AND MANKIND IN WAR AND PEACE

8 POPULAR SCIENCE

Army advertisement in a popular consumer magazine (author's collection).

1947. In addition, the L-15 would be extremely expensive, approximately $30,000 per aircraft, and had not been subjected to Army Ground Forces field tests. Flight tests and service tests were expected to require modifications, resulting in even higher costs. It was also understood that the L-15 aircraft could not be produced in sufficient quantity to replace all Army Ground Forces aircraft and the aircraft assigned to ground units in overseas theaters until 1950.

General Bolte proposed that a commercial model liaison aircraft be procured to replace 50 percent of the Army Ground Forces aircraft during each of the fiscal years 1948 and 1949. It was stated that this action would enable the Air Materiel Command to take the time to develop, test, and modify the XL-15 for AGF use, or, should the XL-15 not prove satisfactory, permit the AMC to develop a new aircraft for AGF use. No additional L-15 aircraft should be procured in quantity until the aircraft had been tested, approved, and production requested by the Army Ground Forces.

In response to this letter, General Eaker, Headquarters Army Air Forces, proposed that 10 YL-15s be procured for service testing in addition to 193 commercial "off-the-shelf" liaison aircraft during each of the fiscal years 1947 and 1948.

The Army Ground Forces concurred with General Eaker's proposal.[6]

This thought pattern had begun with the Wright Field AGF Liaison Officer who, on 6 November 1946, had attended a conference called by the Army Air Forces of light aircraft manufacturers at the Air Materiel Command Headquarters to discuss the purchase of commercial model liaison aircraft in large quantities for immediate issue to the National Guard. He realized that the offered aircraft would cost approximately one-tenth of the price of an L-15 and be procured about ten times faster. This would result in the National Guard having better liaison airplanes for several years than the regular Army.

The outcome of the above actions was the purchase of the Aeronca L-16A for

Aeronca L-16A, 30 of which could be bought for the price of one L-15 (AAHS).

both the U.S. Army and the National Guard at an average unit cost of $1,649 and ten YL-15s for service test at an average unit cost of $51,638 in the 1947 fiscal year.[7]

Engineering Acceptance Inspection of First XL-15 Airplane

In October 1947, an Engineering Acceptance Inspection of the first XL-15 airplane was held at the Boeing Wichita plant. The inspection board was all Wright Field Air Force personnel except for Lt. Col. G. W. LeFever, the AGF Liaison to Wright Field. In addition, the following Army Ground Forces officers were present as observers:

Colonel F. W. Farrell
Lt. Col. C. L. Shepard
Lt. Col. Jack Marinelli
Major D. G. Cogswell
Major L. B. Smith
Major R. L. Long

Some of the above may have had the opportunity to fly the XL-15 during this inspection, but the author has not been able to confirm this.[8]

AGF Flying the XL-15

About the end of January 1948, Army Ground Forces pilots flew the XL-15 to determine whether or not real improvement had been made in the performance and

XL-15 (AAHS).

handling characteristics of the aircraft since the Engineering Acceptance Inspection. They were also to determine if the aircraft was suitable for service test procurement. The pilots involved were

> Lt. Col. Jack Marinelli, AGF Board #1, Fort Bragg, North Carolina
> Major David Cogswell, HQ AGF, Fort Monroe, Virginia
> Major Richard L. Long, AGF Liaison Section, Wright Field, Ohio
> 1st Lt. Paul E. Jackson, AGF

The results of this flight evaluation are not known.[9]

XL-15 Fails to Meet Performance Guarantees

On February 4, 1948, the Air Materiel Command sent a letter to the Chief of Staff, USAF, that the performance tests of the XL-15 had been completed and had been found in several areas not to meet the contract guarantees or the minimum requirements of the "Military Characteristics." The following table details the XL-15 performance as given in that letter.

Table 15.1

Requirement		Min Military Characteristics	Contractor's Guarantee	Actual Performance
Cruising Speed	mph	90	101	95
Service Ceiling	feet	15,000	16,400	12,000
Climb at Sea Level	fpm	600	628	580
Endurance	hours	3.5	4	3.5
Takeoff Distance	feet	600	595	600
Landing Distance	feet	600	517	380
Landing Speed	mph	35	35	24

The letter also reported that Boeing had made all changes that could be made without an excessive cost and delivery time extension. Any further work by the contractor would produce negligible results. The Air Materiel Command felt that the airplane was acceptable, and if the Army Ground Forces agreed, negotiations would be started with Boeing to accept the XL-15 airplanes at a reduced consideration.[10]

On March 17, 1948, the Air Materiel Command was instructed that the above recommendation was acceptable to the Army Ground Forces.[11]

Thus, the first XL-15 airplane was approved for acceptance on March 26, 1948, with a deficiency in performance. The contract value was reduced by $1,005 to compensate for this reduction in performance. The total amount paid to Boeing for the XL-15 program was $378,043.15.[12]

On July 12, 1948, after completing Air Force flight tests, XL-15 46–520 was released to the Wright Field Aircraft Laboratory for static testing of the airframe.

YL-15 s for Service Test

On 28 April 1948, The USAF Deputy Chief of Staff for Material informed the Air Materiel Command that the Army Field Forces (AFF) (the new name for AGF) had advised the USAF Headquarters that they realized the short field performance of the aircraft would be seriously affected by the limited service ceiling and reduced rate of climb, but that they agreed there would be no benefit to be derived from further modification of the aircraft short of redesign. Accordingly, the Army Field Forces considered that the aircraft with its present performance was acceptable for service testing, but the decision as to the ultimate suitability must be deferred pending completion of the necessary service test. The Air Materiel Command was advised to continue the procurement of YL-15 aircraft for service testing. Ten YL-15s were ordered at an average unit cost of $51,638.[13]

The revised delivery schedule for the YL-15 as of November 5, 1948, was for one airplane to be delivered in November 1948 and the remaining nine in December. The Air Force also on that date stated that they did not believe that the Army Field Forces would be satisfied with the performance of the YL-15 in their upcoming service test and that the re-engining of the XL-15 with a 140 horsepower Lycoming engine should be done with priority at Wright Field at USAF expense as the Army did not have funds to do this.[14]

On December 22, 1948, the second XL-15 (sn 46–521) and the first YL-15 (sn 47–423) were flown from Boeing Wichita to Wright Field for testing by the Air Force. The other nine YL-15s, after completion, went to Fort Bragg, North Carolina, for service testing by the Army Field Forces. The last delivery from Boeing to the Army was made in March 1949.[15]

XL15 with 140 HP Lycoming O-290-A2 (NARA via AEHS).

XL-15 46–521, flown by an Army crew, departed Wright-Patterson AFB on September 14, 1949, to Fort Bragg. In February 1950, the Army loaned it back to the USAF to conduct lateral and directional control evaluations after it had been modified by Boeing.

On November 21, 1950, YL-15 47–423 was transferred from the USAF to the U.S. Army. The Army now had all ten YL-15s.

Top: YL-15s at Boeing ready for delivery (author's collection). *Bottom:* YL-15s before departure to Fort Bragg (author's collection).

Most of the service test flying was at Fort Bragg, North Carolina, but also winter testing was done in Alaska, desert testing in Arizona, and high-altitude field testing at Leadville, Colorado. The L-15 was displayed along with other aircraft and weapons when President Truman visited Fort Bragg on October 4, 1949, to see a demonstration of the Army in action.[16]

The author has not found any documents giving the results of the Army Field Forces' YL-15 service test, but it appears that the Army was not very happy with the YL-15 because in late 1949, while the service test was still going on, the Army once again called for a competition to select an off-the-shelf liaison airplane.

L-19 Fly-off Competition

This was a way for the Army to be able to get around having the Air Force select another aircraft for them. Even though the Air Force would officially administer the competition, the Army would write the rules and do the judging. The winner would be selected in a fly-off competition. Several light plane companies competed in the fly-off April 6–14, 1950, at Wright Field. For the most part, these were not stock civilian airplanes but were new designs explicitly built for this competition. They, however, often used parts from existing production aircraft.

The Cessna Model 305 was announced the winner on May 29, 1950, with an order for 418 airplanes to be designated the L-19A. At the time of the competition start, Cessna had an engineering department of only 18 people who were already working on several commercial projects. They assigned a handful of their engineers to start the design of the Model 305 in August 1949. Their entry used the tail of a

Cessna L-19A, the airplane the Army bought in quantity instead of the L-15 (AAHS—Chuck Stewart Collection).

YL 15 at Minto, Alaska, in April 1954 on frozen river (U.S. Fish & Wildlife Service).

Model 195 and the wings of a Model 170 with new flap hinges to allow 60° flap deflection. A new fuselage with tandem seating and lots of visibility was drawn up full size. The engine was a Continental O-290–11 with a 213 horsepower takeoff rating. Just weeks after the design began, construction was started. The prototype building went from September 8 to December 8, 1949. The first flight was on December 14, 1949. The first production aircraft was delivered in late 1950, and the first L-19 reached a combat unit in Korea on February 16, 1951. The L-19 had better performance, was developed in one-quarter of the time, and cost only 20 percent as much as the L-15. The Cessna L-19 was designed and constructed in a small plant that was in sight of the huge Boeing Wichita plant.[17] The Boeing L-15 was dead, and the Army would shortly transfer the YL-15s to the U.S. Fish and Wildlife Service in Alaska.

But as seen in the next chapter, the Boeing L-15 team did not give up and kept trying.

16

Boeing's Projected Follow-on L-15 Versions

As mentioned in previous chapters, Boeing had on numerous occasions proposed to the Air Force versions of the L-15 with alternate engines. The USAF was interested in several of these proposals and actually modified the second XL-15 with a 140 HP Lycoming O-290-A2 engine. Results with this higher power engine were disappointing.[1]

Army L-19 Fly-off Competition

In 1949 Boeing had entertained submitting a redesigned version of the L-15 into the L-19 fly-off competition. The Boeing entry would have been a modified YL-15 design with a three-foot reduction in wingspan, an increased fuel load, and a new engine of 190 horsepower.

Boeing decided not to enter the L-19 competition as it would have required building a new aircraft at company expense.[2]

Unsolicited Liaison Aircraft Proposal

In July 1950, Project Engineer Earl Weining proposed to Boeing Wichita management that an unsolicited liaison aircraft proposal be made to the U.S. Army based on a highly redesigned L-15. It was Weining's opinion that because of the Korean War, the Army would need more aircraft than Cessna could produce.

The existing YL-15 rectangular wing would be replaced with a new 175-square-foot tapered wing. The new wing would have a root chord and airfoil the same as the YL-15, but tapering to a 45-inch USA-35B airfoil at the tip. The trailing edge would be upswept and normal to the aircraft centerline, with the leading edge being swept back. The 40-foot wingspan would be retained. The external flap would be replaced with a NACA slotted flap with an area of at least 50 square feet. The spoiler span would be doubled, and very small feeler ailerons of 2.7 square feet would be at each tip.

The tail boom would be rotated downward around the two lower front attach points to the fuselage body, so the last boom bulkhead would be lowered vertically 32 inches. New rear observer's doors would be fitted with the upper rear point of

A Boeing L-15 Scout serves as "eyes" for an Infantry reconnaissance detail.

Sharp Eyes for Our Ground Forces

The Boeing L-15—newest Army Ground Forces liaison plane—marks a long stride beyond the valiant "puddle jumpers" that served the Infantry, Artillery and Armored Cavalry in World War II.

The L-15 was designed by Boeing to meet U. S. Air Force standards and Army Ground Forces requirements. Like all Boeing planes, commercial and military, it is designed for a specific job, with maximum utility as a primary objective.

Revolutionary in appearance and performance, it can hop in and out of a small meadow without difficulty. It can cruise at about 100 mph, land at less than 35.

For safety of its crew the L-15 has armor-plate protection, blind flight instruments, self-sealing tanks, and dual cable control system. The unique gondola design with full swiveling observer's seat provides full vision—up and down, forward, backward and to either side. Powerful two-way radio maintains constant touch with the ground.

Designed for use under all conditions, from tropical to arctic, the L-15 may be operated with wheels, skis or floats. The whole airplane can be "folded up" for towing behind a jeep or loading aboard a 2½-ton standard army truck.

The L-15 is an example of effective cooperation between the Armed Forces and the aircraft industry. It was engineered and is being manufactured by the Boeing Wichita Division, which produced for the Air Force and the Navy, during the war, more primary training planes than any other plant.

A "folded" L-15 can be towed through narrow, winding woods roads.

For the Air Force, Boeing is building the B-50 bomber, XB-47 jet bomber and C-97 transport; for the Army Ground Forces, the L-15 Scout liaison plane; and for six major airlines, the twin-deck Boeing Stratocruiser.

BOEING

Boeing 1948 magazine ad for the L-15 (author's collection).

16. Boeing's Projected Follow-on L-15 Versions

Earl Weining's redesigned L-15 proposal to Boeing management (author's collection).

the doors moved to the rear of the boom transition point. The pilot's door would be redesigned to hinge at the front.

The twin vertical tails would be replaced by a 16.5-square-foot single vertical tail mounted on the top of the boom. The tip of the vertical would not be more than 95 inches above the three-point ground line. The existing YL-15 stabilizer and elevators would be used with tips added.

The YL-15 engine would be used, but the engine nose bug inlet area would be reduced 20 percent, and the exit opening area increased 25 percent. A constant-speed metal propeller would be used.

At the time, Boeing Wichita was busy setting up production for the B-47 Stratojet and had little need or manpower to pursue the L-15 program.

The following table presents the performance of the modified YL-15 design and that of Weining's proposed L-15 to the requirements and those of the winning Cessna L-19 aircraft.[3]

Table 16.1. L-15 Follow-on Studies—1950

Requirement		Spec Requirement	Modified YL-15	Proposed L-15	Cessna L-19A
Weight	lbs		2,400	1,980	2,100
Power	HP		190	125	213
Span	ft	37 max	37	40	36
TO dist over 50 ft—sod	ft	300–600	500	530	580
Landing dist over 50 ft—sod	ft	300–600	450	450	550

Requirement		Spec Requirement	Modified YL-15	Proposed L-15	Cessna L-19A
Landing Speed	mph		40	40	
Stall Speed	mph		42.5	42.5	48
Max R/C	fpm	800–1,000	1,050	867	1,290
Service Ceiling	ft	16,000–18,000	17,000	19,500	18,500
Cruise	mph	90–125	110	110	117
Endurance	hours	3–4	3.5	3.3	

Utility Airplane

In the early 1950s, Earl Weining tried to sell Boeing on a utility airplane project based on the L-15. The potential uses of this aircraft were short-haul transportation of passenger and freight, crop dusting, seeding, spraying, fertilizing, general farm and range service, harvest combine service, survey, pipeline inspection, police work, ambulance service, forestry patrol, wire laying, prospecting, photography, construction engineering work, and military liaison. It was to be a true utility or "station wagon" airplane that could pay its way. It was felt there would be a large market for this aircraft not just in North America but also throughout South America, which had few good roads, highways, or railroads.

Weining stated that 72,000 engineering manhours and $9,000 of wind tunnel testing had gone into the L-15 design. Basing the new utility airplane on this previous work would allow the new aircraft to be designed, built, tested, and CAA certified in ten months with an additional expense of 11,000 engineering man-hours and $2,000 of wind tunnel testing. He also argued that the 12-man L-15 engineering team was in place and could jump on this immediately.

Boeing management was not convinced, and the L-15 died in place.[4]

The Outcomes of the L-15

The U.S. Army did not accept the Boeing L-15 as a production liaison airplane for many reasons. It did not meet performance guarantees and requirements. Boeing's development time was many times longer than what the Army expected. The L-15's unit cost was magnitudes above that of other comparable or better liaison aircraft. Its flying characteristics were nonstandard and different from other liaison airplanes. The airplane was designed for depot maintenance with replaceable parts, while the Army expected an aircraft to be operated, maintained, and repaired under front-line field conditions. And finally, the Army probably did not appreciate how little input the Air Force gave them into the program. Hopefully, it taught the Air Force something about working with its sister armed forces.

But what about the positive things that came out of the L-15 project? The most significant plus was that it enabled the Boeing Wichita division to stay in operation and be available when needed for the later B-47 Stratojet production program. It gave many Boeing engineers experience with light aircraft, which they would use

U.S. Fish and Wildlife Service YL-15 in Alaska (U.S. Fish and Wildlife Service).

in later years working at Beech, Cessna, Lear, and other general aviation companies. It gave the Army understanding and knowledge that they needed to have a prominent role in developing and purchasing aircraft for Army use. It allowed the U.S. Fish and Wildlife Service, which had little money budgeted for aviation, to obtain aircraft.

Appendix: L-15 Design Patent

Patented Apr. 12, 1949

Des. 153,411

UNITED STATES PATENT OFFICE

153,411
DESIGN FOR AN AIRPLANE

Harold W. Zipp, Verne L. Hudson, and Earl O. Weining, Wichita, Kans., assignors to Boeing Airplane Company, Wichita, Kans.

Application July 24, 1947, Serial No. 140,486

Term of patent 14 years

(Cl. D71—1)

To all whom it may concern:

Be it known that we, Harold W. Zipp, Verne L. Hudson, and Earl O. Weining, all citizens of the United States, residing at Wichita, Kansas, have jointly invented a new, original, and ornamental Design for an Airplane, of which the following is a specification, reference being had to the accompanying drawings forming a part thereof.

Figure 1 is a top plan view of an airplane embodying the new design;

Figure 2 is a front view of the same airplane;

Figure 3 is a side elevation of the same airplane; and

Figure 4 is a three-quarters front perspective view of the same airplane.

Figure 5 is a fragmentary rear perspective view of the fuselage, the tail boom and empennage having been omitted for better disclosure of the window section at the rear of the fuselage.

We claim:

The ornamental design for an airplane, substantially as shown.

HAROLD W. ZIPP.
VERNE L. HUDSON.
EARL O. WEINING.

REFERENCES CITED

The following references are of record in the file of this patent:

UNITED STATES PATENTS

Number	Name	Date
D. 55,504	Curtiss	June 22, 1920
D. 106,939	Waterman	Nov. 9, 1937
D. 133,777	Gardenhire	Sept. 15, 1942
D. 136,635	Fontaine	Nov. 9, 1943

OTHER REFERENCES

Jane's All The World's Aircraft, 1941 edition, page 228C, figure—The Vultee O–49 "Vigilant."

238   Appendix: L-15 Design Patent

April 12, 1949.   H. W. ZIPP ET AL   **Des. 153,411**

AIRPLANE

Filed July 24, 1947   2 Sheets–Sheet 1

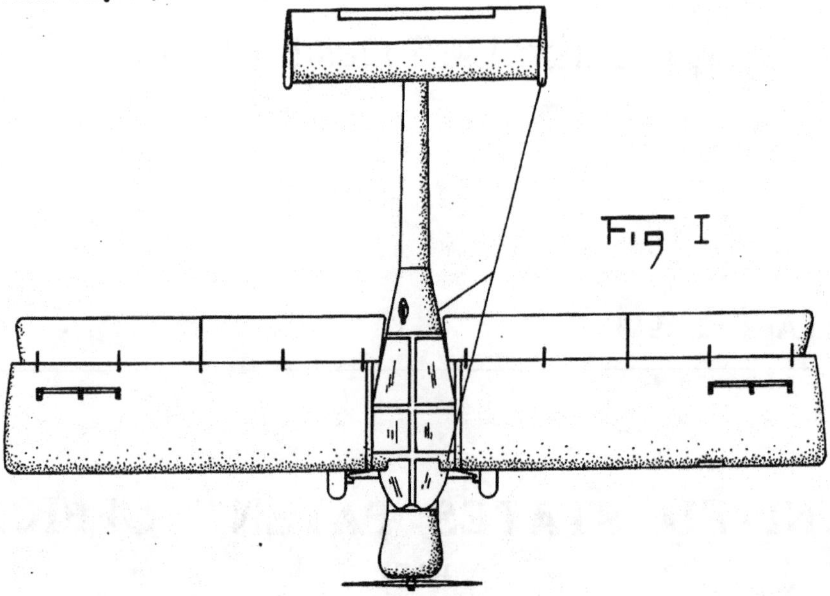

Fig 1

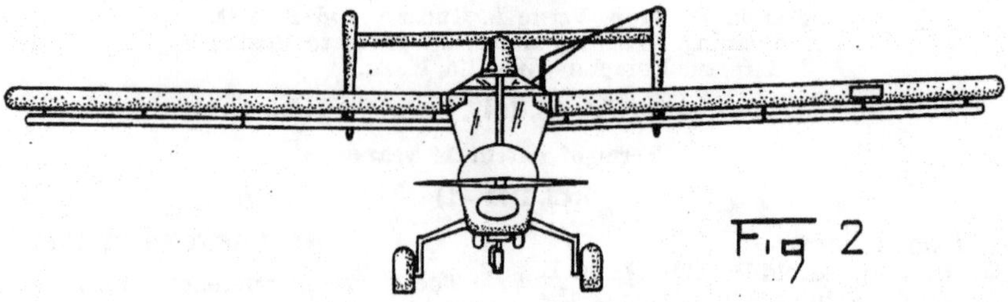

Fig 2

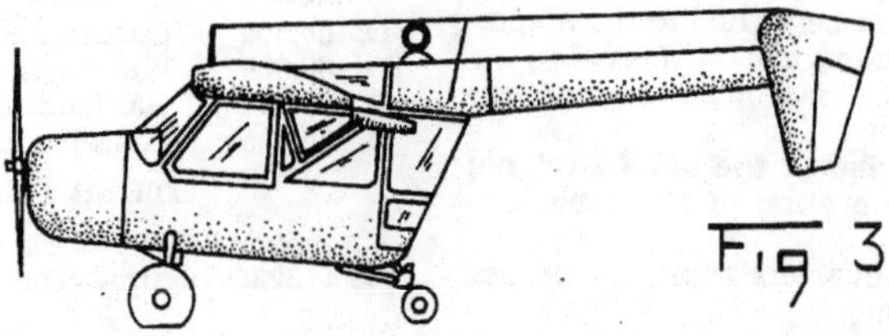

Fig 3

HAROLD W. ZIPP
VERNE L. HUDSON
EARL O. WEINING
INVENTORS

BY *Hubert Miller*
ATTORNEY

Appendix: L-15 Design Patent 239

April 12, 1949. H. W. ZIPP ET AL **Des. 153,411**
AIRPLANE

Filed July 24, 1947 2 Sheets—Sheet 2

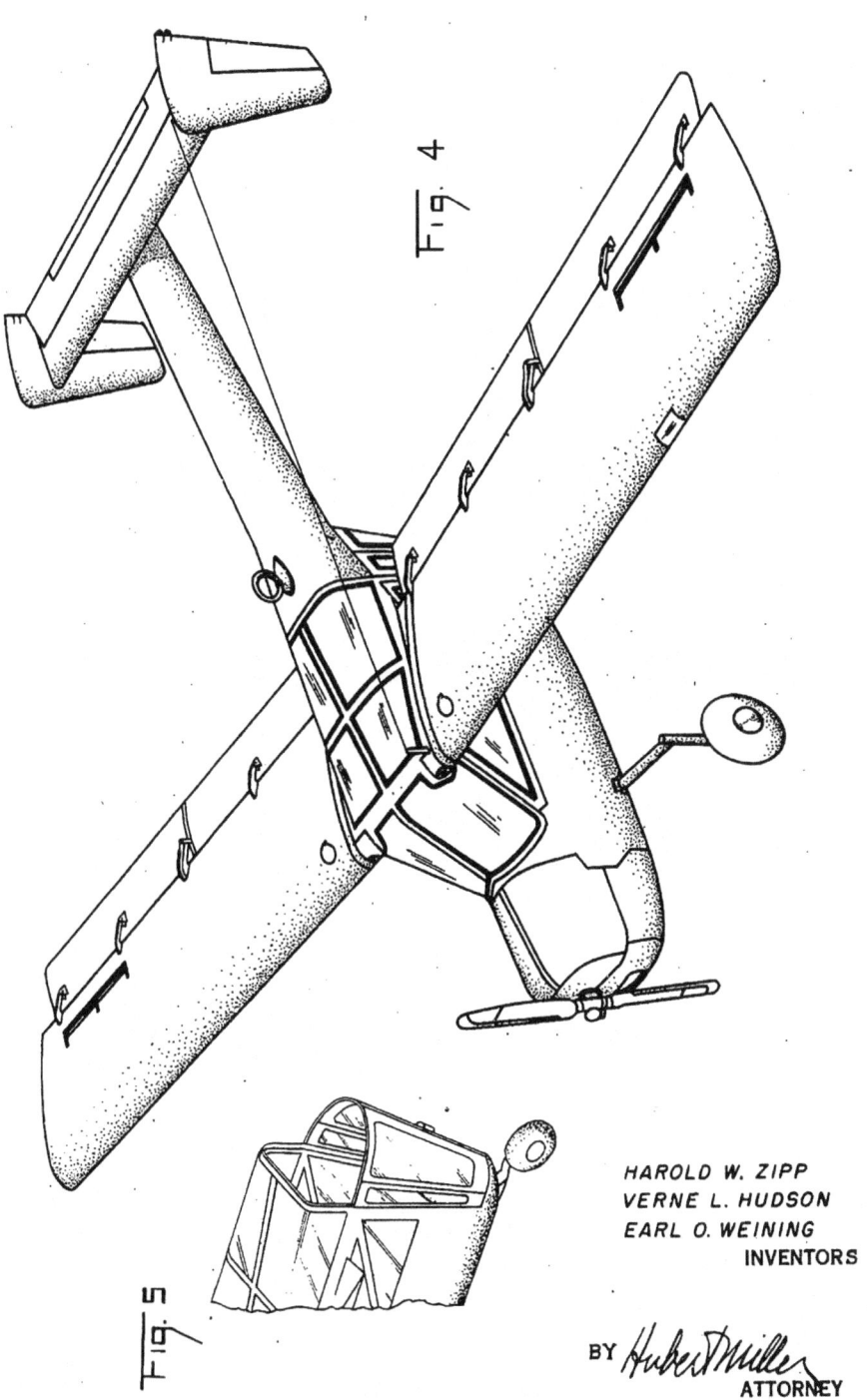

HAROLD W. ZIPP
VERNE L. HUDSON
EARL O. WEINING
INVENTORS

BY *Hubert Miller*
ATTORNEY

Chapter Notes

Chapter 1

1. *Civil Aviation and the National Economy* (Washington, D.C.: United States Dept. of Commerce, Civil Aeronautics Administration, Sept 1945), ix–x, 2–3, 40–43, 97–99.
2. *Civil Aviation*, vi–x, 47–48.
3. *Civil Aviation*, 37, 95-102.
4. *Civil Aviation*, 99.
5. Joseph P. Juptner, *U.S. Civil Aircraft, Volume 8* (Fallbrook, CA: Aero Publishers Inc., 1980), 241–244; and Rod Simpson, *The General Aviation Handbook* (Hinckley, England: Midland Publishing, 2005), 249–250.
6. Juptner, 283–284; and Ray Wagner, *Mustang Designer* (New York: Orion Books, 1990), 166–167.
7. Rene J. Francillon, *McDonnell Douglas Aircraft Since 1920: Volume I* (Annapolis, MD: Naval Institute Press, 1988), 410–411.
8. Rene J. Francillon, *Lockheed Aircraft Since 1913* (Annapolis, MD: Naval Institute Press, 1988), 259–260, 281–283.
9. Randy Mertens, *Closet Cases* (Kansas City, MO: Pilot News Press, 1980), 59–61.
10. Mertens, 61–63.
11. George Spatt, "Wing on a Pivot," *Flying*, Vol. 38, No. 5, May 1946, 42–44, 112.
12. Wikipedia contributors, "Convair Model 118," *Wikipedia, The Free Encyclopedia,* https://en.wikipedia.org/wiki/Convair_Model_118 (accessed May 17, 2017); Peter M. Bowers, *Unconventional Aircraft* (Blue Ridge Summit, PA: Tab Books Inc., 1984), 196–197.

Chapter 2

1. E. O. Weining, "External Airfoil Flaps as a High-Lift Device" (29 Nov 1948), 1.
2. James M. Wickham, "Research in Simplified Control" (read by D. C. Heinburger to the Wichita Chapter Society of Automotive Engineers, 20 January 1949), 1.
3. Wickham, 2.
4. Wickham, 2.
5. Wickham, 2–3.
6. Wickham, 3; I do not know if Wickham had a pilot's license at the time he made this negative comment, but in later years he would be very active in general aviation and even design and build six homebuilt airplanes.
7. Wickham, 3–4.
8. Wickham, 4.
9. Wickham, 5.
10. Wickham, 6.
11. Wickham, 5.
12. George Spatt, "Wing on a Pivot," *Flying*, Vol. 38, No. 5, May 1946, 42–44, 112.
13. Wickham, 6.
14. Wickham, 6–7.
15. Wickham, 7.
16. Weining, 1; Wickham, 7.
17. Weining, 1; Wickham, 7–8.
18. Wickham, 8.
19. Wickham, 8.
20. Wickham, 8.
21. Wickham, 8–9.
22. Wickham, 9.
23. Wickham, 9.
24. Wickham, 9.
25. Wickham, 9–10.
26. Wickham, 10.
27. Weining, 2; Wickham, 10–11.
28. Weining, 3; Wickham, 11.
29. Wickham, 11.
30. Wickham, 11.
31. Wickham, 11–12.
32. Wickham, 12.
33. Weining, 2; Wickham, 12.
34. Weining, 2.
35. Weining, 3; Wickham,12–13.
36. Wickham, 13.
37. Wickham, 13.
38. Wickham, 14.
39. Wickham, 13–14.
40. Wickham, 14.
41. Weining, 3.
42. Wickham, 14.
43. Wickham, 14.

Chapter 3

1. R. HB. Davis, Major, Air Corps, Acting Chief of Administration, Air Technical Service Command, Wright Field, Dayton, OH, "Field Artillery Observation Aircraft," Technical Instructions

TI-2133, Addendum No. 3 to Engineering Division, 27 September 1945.

2. George E. Price, Col, Air Corps, Chief Aircraft Projects Section, Service Engineering Subdivision, Engineering Division, Air Technical Service Command, Wright Field, Dayton, OH, "Field Artillery Observation Aircraft—Request for Bids," letter to 26 aircraft manufacturers, 4 February 1946.

3. *Method of Evaluation for Aircraft, Observation (Field Artillery)* Inclosure No. 3 of RFB

4. L. V. Cook, *XL-15 Liaison Airplane, "Scout": Final Report of Development, Procurement, Inspection, Testing, and Acceptance,* AF Technical Report No. 6046 (Wright-Patterson Air Force Base, Dayton, OH: USAF Air Materiel Command, July 1949), 1.

5. *Type Specification for Land Observation Airplane)* (Bellanca Aircraft Corporation Specification A-2, February 28, 1946); *Type Specification for Land Observation Airplane* (Bellanca Aircraft Corporation Specification A-3, February 28, 1946); and M.C. Wardle, *Aerodynamics Report, Preliminary Design of Liaison Observation Airplane* (Bellanca Aircraft Corporation Report No. 469, March 4, 1946).

6. Harold W. Zipp, Chief Engineer, Wichita Division, Boeing Airplane Company, "AAF Liaison Airplane—Proposed Design For," letter to Commanding General, Air Technical Service Command, Wright Field, Dayton, OH, Attn: Engineering Division—Brig. Gen L.C. Craigie, 5 December 1945.

7. *Model 451-2 Preliminary Performance, Stability and Control Characteristics* (Boeing-Wichita Report No. WD-12056, 6 March 1946); and E.O. Weining, *Development of the Model L-15 Liaison Airplane and Design Development of Liaison Aircraft* (Boeing-Wichita Report No. WD-13024, 20 June 1950).

8. *Proposal for Army Air Forces Observation Airplane* (Consolidated Vultee Aircraft Corp., San Diego, CA, March 6, 1946); W.F. Brown, *Preliminary Model Specification for Army Air Forces Observation Airplane (Field Artillery)One Lycoming O-290-7 Engine* (Consolidated Vultee Aircraft Corp Report ZD-016, San Diego, CA, 25 February 1946); and C. Blake, F. Leward, and K. Smith, *Aerodynamic Report for Army Air Forces Observation Airplane* (Consolidated Vultee Aircraft Corp Report ZA-152, San Diego, CA, March 1, 1946).

9. "Ludington-Griswold," http://aerofiles.com/_lo.html (accessed November 13, 2018); and *Proposal for the Development of a Field Artillery Observation Airplane* (Ludington-Griswold, Inc., Saybrook, CT, 6 March 1946).

10. *Piper Model PA-9 Engineering Data* (Piper Aircraft Corp. Report No. 504, Lock Haven, PA, 3/1/1946).

11. *Discussion of Engineering and Suitability Characteristics—Manufacturer's Ratings,* Inclosure No. 2 of following referenced letter.

12. Commanding General, Air Materiel Command, "Informal Design Competition—Field Artillery Observation Aircraft (Liaison Type)," letter to Commanding General. Army Air Forces, Washington, D.C., 18 April 1946.

13. J. E. Schaefer, Vice President, Wichita Division, Boeing Airplane Company, "Field Artillery Observation Aircraft—Request for Bids," letter to Commanding General, Air Technical Service Command, Wright Field, Dayton, OH, 4 March 1946.

14. R. K. Taylor, Brig. Gen., USA, Air Materiel Command, "Procurement of Observation Aircraft (Liaison Type)," Technical Instructions to Engineering Division, 6 May 1946.

15. AF TR 6046, 1–2.

Chapter 4

1. Harold W. Zipp, Chief Engineer, Wichita Division, Boeing Airplane Company, "AAF Liaison Airplane—Proposed Design For," letter to Commanding General, Air Technical Service Command, Wright Field, Dayton, OH, Attn: Engineering Division—Brig. Gen L.C. Craigie, 5 December 1945.

2. *Model 451–2 Preliminary Performance, Stability and Control Characteristics* (Boeing-Wichita Report No. WD-12056, 6 March 1946); and E. O. Weining, *Development of the Model L-15 Liaison Airplane and Design Development of Liaison Aircraft* (Boeing-Wichita Report No. WD-13024, 20 June 1950).

3. WD-12056, 7, 18–19, 54, 57.
4. WD-12056, 7, 18–19, 42.
5. WD-12056, 7, 18–19, 51.
6. WD-12056, 7.
7. WD-12056, 29.
8. WD-12056, 29.
9. WD-12056, 7.
10. WD-12056, 17.
11. WD-12056, 4.
12. WD-12056, 6.
13. WD-12056, 20, 46, 106.
14. WD-12056, 47, 115, 202.
15. WD-13024, 6.

Chapter 5

1. H. M. Basow and B. C. Hainline. *Development Tests of Beaded Wing Skin* (Boeing-Wichita Report No. WD-10541, August 27, 1946).

2. E. O. Weining, *Development of the Model L-15 Liaison Airplane and Design Development of Liaison Aircraft* (Boeing-Wichita Report No. WD-13024, 20 June 1950), 6.

3. E. O. Weining, "External Airfoil Flaps as a High-Lift Device" (29 Nov 1948), 5

4. C. M. Seibel, *Model XL-15 Estimated Performance, Stability and Control Characteristics* (Boeing-Wichita Report No. WD-12130, 12 June 1947), 11–12, 88–89.

5. WD-12130, 113–114.
6. "Rudder Control," undated note in Weining personal file.
7. WD-12130, 38.
8. WD-12130, 39.
9. "External Airfoil Flaps," 3.
10. Harold W. Zipp, Chief Engineer, Wichita Division, Boeing Airplane Company, "Contract W33–038 ac-15054—Model XL-15 Airplane—Propeller for," letter to Commanding General, Air Materiel Command, Wright Field, Dayton, OH, Attn: TSEOA-5, 18 March 1947.
11. "External Airfoil Flaps," 1–3.
12. J. E. De Remer, *Memorandum Report on Mock-Up Inspection of Xl-15 Airplane at Boeing Aircraft Company, Wichita, Kansas on 5 September 1946.* TSEPP-436-14 (Wright Field, Dayton, OH: Power Plant Laboratory, Engineering Division, Air Materiel Command, Army Air Forces, 18 Oct 1946); and L. V. Cook, *XL-15 Liaison Airplane, "Scout": Final Report of Development, Procurement, Inspection, Testing, and Acceptance,* AF Technical Report No. 6046 (Wright-Patterson Air Force Base, Dayton, OH: USAF Air Materiel Command, July 1949), 2.
13. AF TR 6046, 2.
14. E. H. Rowley, *Initial Ground and Flight Tests on XL-15.*
15. E. H. Rowley, *Time Before Space: An Airman's Odyssey—from Biplanes to Rockets* (Manhattan, KS: Sunflower University Press, 1994), 173.
16. *Initial Ground and Flight Tests on XL-15.*
17. E.O. Weining, *Summary Analysis of Phase I Flight Tests Model XL-15 Airplane* (Boeing Wichita Report No. WD-12129, 8 April 1948), 5.
18. *Time Before Space,* 174.
19. Harold W. Zipp, Chief Engineer, Wichita Division, Boeing Airplane Company, "Contract W33–038 ac-15054—Model XL-15 Airplane—Extension of Contract Delivery Date (Change No. 19) and Request for Additional Flight Test Time (Change No. 20)," letter to Commanding General, Air Materiel Command, Wright Field, Dayton, OH, Attn: TSEOA-5, 20 July 1947.
20. WD-12129, 5–6; Miller, Manley, and Morley, *Control Surface Analysis for Model XL-15 & YL-15* (Boeing-Wichita Report No. WD-10669 Rev2, December 15, 1948), D4, E2, E7.
21. WD-12129, 6.
22. *Time Before Space,* 173.
23. Harold W. Zipp, Chief Engineer, Wichita Division, Boeing Airplane Company, "Contract W33–038 ac-15054—Model XL-15 Airplane—Accident Report on," letter to Commanding General, Air Materiel Command, Wright Field, Dayton, OH, Attn: TSEOA-5, 13 August 1947.
24. E.O. Weining, *Flight Test of Carburetor with Automatic Altitude Compensation Model Xl-15* (Boeing Wichita Report No. WD-12254, 24 February 1948).
25. WD-12129, 6–7.
26. "External Airfoil Flaps," 9–10.
27. AFTR 6046, 3.
28. Wickham, Rowley, and Weining, *Summary of Phase III Flight Tests Model XL-15 Airplane* (Boeing Wichita Report No. WD-12303, 17 December 1948), 4–5.
29. WD-13024, 10–11, 35–36; "External Airfoil Flaps," 10–11.
30. WD-12303, 5–13.
31. "External Airfoil Flaps," 11.
32. WD-12303, 14.
33. WD-12303, 14–15.
34. WD-12303, 16–17.
35. WD-12303, 18–19.
36. *Time Before Space,* 202.
37. AF TR 6046, 3.

Chapter 6

1. George E. Price, Col, Air Corps, Chief Aircraft Projects Section, Service Engineering Subdivision, Engineering Division, Air Technical Service Command, Wright Field, Dayton, OH, "Field Artillery Observation Aircraft—Request for Bids," letter to 26 aircraft manufacturers, 4 February 1946.
2. *Model 451–2 Preliminary Performance, Stability and Control Characteristics* (Boeing-Wichita Report No. WD-12056, 6 March 1946).
3. *Airplane, Observation (Field Artillery) Construction Specification for Basic Flight Test Model XL-15, Boeing-Wichita Model XL-15, Airplane Serial No. 46–520* (Boeing Wichita Report No. WD-12102, 1 October 1946).
4. C. M.Seibel, *Model XL-15 Estimated Performance, Stability and Control Characteristics.* (Boeing-Wichita Report No. WD-12130, 12 June 1947).
5. Charles Seibel worked days at Boeing and in his time off was designing a helicopter. He would go on to form Seibel Helicopter Company, which was later sold to Cessna.
6. *Pilot's Handbook Flight Operating Instructions for XL-15 Airplane Serial Number 46–521.* (Boeing Wichita Document WD-12174).
7. Charles H. La Fountain, *Memorandum Report on Flight Tests of the XL-15, USAF No. 46–520.* MCRFT-2120 (Wright-Patterson Air Force Base, Dayton, OH: Air Materiel Command, 5 January 1948).
8. *Time Before Space,* 174–176.
9. MCRFT-2120.
10. Charles H. La Fountain and Robert H. Shaefer, LTC, USAF, *Memorandum Report on Flight Tests of the XL-15, USAF No. 46–520.* MCRFT-2120 (Addendum 1) (Wright-Patterson Air Force Base, Dayton, OH: Air Materiel Command, 6 February 1948).
11. Thomas J. Borgstrom and Robert L. Northup, 1st Lt, USAF, *Memorandum Report on Comparative Propeller, Engine Cooling and Longitudinal Stability and Control on the XL-15, USAF No. 46–520,* MCRFT-2152 (Wright-Patterson Air Force Base, Dayton, OH: Air Materiel Command, 20 July 1948).

12. L. V. Cook, *XL-15 Liaison Airplane, "Scout": Final Report of Development, Procurement, Inspection, Testing, and Acceptance*, AF Technical Report No. 6046 (Wright-Patterson Air Force Base, Dayton, OH: USAF Air Materiel Command, July 1949), 3.

13. Aircraft Project and Historical File 1942–1951 YL-15–47–423, RG 342, NARA II

14. E. O. Weining, *Airplane Characteristics Summary Model YL-15* (Boeing Wichita Report No. WD-12280, 27 May 1948).

15. *Pilot's Handbook Flight Operating Instructions for YL-15 Airplane Serial Numbers 47–423 Through 47–432*. (AN 01–20LAA-1, Nov 1, 1948).

16. D.C. Heimburger, *Pilot Comments and Flight Performance Data on the YL-15 Airplane*. (Boeing-Wichita Report No. WD-12827, 10 Jan 1949).

17. *Handbook Flight Operating Instructions USAF Model YL-15 Aircraft* (AN 01–20LAA-1, 23 March 1949).

18. James M. Wickham, *YL-15 Standard Characteristics* (Boeing Wichita Report No. WD-12366, 15 July 1949).

19. E. O. Weining, *Development of the Model L-15 Liaison Airplane and Design Development of Liaison Aircraft* (Boeing-Wichita Report No. WD-13024, 20 June 1950).

20. Harold W. Zipp, Chief Engineer, Wichita Division, Boeing Airplane Company, "Contract W33–038 ac-15054—Model XL-15 Airplane Fuel Requirements, Possible Engine Development Resulting from Change in," letter to Mr. J. B. Risbel, Chief Aircraft Project Section, Service Engineering Subdivision, Engineering Division, Air Materiel Command, Wright Field, Dayton, OH, 15 August 1946.

21. J. E. Schaefer, Vice President, Wichita Division, Boeing Airplane Company, "Contract W33–038 ac-15054, Model XL-15 Airplane—Alternate Engine Installation," letter to Commanding General Air Materiel Command, Wright Field, Dayton, OH, 7 July 1947.

22. J. E. Schaefer, Vice President, Wichita Division, Boeing Airplane Company, "Contract W33–038 ac-16945, Model YL-15 Airplanes—Proposal for Engine Change," letter to Commanding General, Air Materiel Command, Wright Field, Dayton, OH, 15 October 1947; George E. Schaetzel, Col, Air Corps, Chief, Aircraft Section, Procurement Division, Air Materiel Command, "W33–038 ac-16945, Model YL-15 Aircraft Revision of Planning," letter to Chief of Staff U.S. Air Force, Washington, D.C., 24 October 1947.

23. George E. Schaetzel, Col, USAF, Chief, Aircraft Section, Procurement Division, Air Materiel Command, "W33–038 ac-16945, Model YL-15 Aircraft Revision of Planning" letter to USAF Inspector in Charge, Boeing Airplane Company, Wichita Division, Wichita, KS, 5 December 1947.

24. George E. Schaetzel, Col, USAF, Chief, Aircraft & Missiles Section, Engineering Division, "Installation of 140 H.P. Engine in XL-15 Airplane" conference minutes 5 Nov 1948; and R. T. Vandegrift, Service Manager, Lycoming Division, Avco Manufacturing Corp, Williamsport, PA, "Loan of O-290-A2 Engine for XL-15 Airplane" letter to Commanding General, Air Materiel Command, Wright Patterson AFB, Dayton OH, Attn: MCREOA, December 14, 1948.

25. Jack H. King and Thomas J. Cecil, Capt, USAF, *Memorandum Report on Performance Tests of the XL-15 Airplane, USAF No. 46–521 with a Lycoming O-290-A2 140 Horsepower Engine Installed*, MCRFT-2243 (Wright-Patterson Air Force Base, Dayton, OH: Air Materiel Command, 1 September 1949).

26. MCRFT-2243, 3; The data reduction methods used by AMC were the same the author used at Beech and Piper for small aircraft.

27. MCRFT-2243, 4; Since there was a 15 brake horsepower increase, this would indicate that the propeller used was not the proper one for the increased power.

Chapter 7

1. *Airplane, Observation (Field Artillery) Construction Specification for Basic Flight Test Model XL-15, Boeing-Wichita Model 451, Contract W33–038 AC-15054, Airplane Serial No. 46–520* (Boeing Wichita Report No. WD-12102A, 29 August 1949), Appendix III p-4.

2. WD-12102A, 1.

3. N. Miller, *Wing Stress Analysis—Volume II for Models XL-15 & YL-15* (Boeing-Wichita Report No. WD-10666 Rev2, December 15, 1948), B7-B9, B13-B18; WD-12102A, 19; and *Handbook Erection and Maintenance Instructions USAF Model YL-15 Aircraft*. AN 01–20LAA-2 (10 January 1949), 19.

4. WD-10666, E-44; WD-12102A, 20; and AN01–20LAA-2, 20; Miller, Manley, and Morley, *Control Surface Analysis for Model XL-15 & YL-15* (Boeing-Wichita Report No. WD-10669 Rev2, December 15, 1948), 5, A3-A5, A8.

5. WD-10669, 5, B2-B3; and AN01–20LAA-2, 22.

6. WD-10669, 5, C3; WD-12102A, 21; and AN01–20LAA-2, 23–24.

7. WD-10669, C3; WD-12102A, 21; and AN01–20LAA-2, 24–26.

8. WD-10669, 5, D3-D4; WD-12102A, 21; and AN01–20LAA-2, 27.

9. WD-10669, D3-D4; WD-12102A, 21–22; and AN01–20LAA-2, 28.

10. Wayne Dalrymple, *Organizational Requirements and Design Criteria Model XL-15*. (Boeing Wichita Report No. WD-12110 Rev3, 14 March 1947), 40–41; WD-12102A p22; and AN01–20LAA-2, 43, 46–47.

11. *Handbook Flight Operating Instructions USAF Model YL-15 Aircraft* (AN 01–20LAA-1, 23 March 1949), 4.

12. AN01–20LAA-2, 93; and AN01–20LAA-1 (1949), 9.

13. AN01–20LAA-2, 93.
14. AN01–20LAA-2, 93.
15. AN01–20LAA-1 (1949), 5, 9.
16. AN01–20LAA-2, 95.
17. AN01–20LAA-2, 95.
18. AN01–20LAA-2, 95; and AN01–20LAA-1 (1949), 25.
19. AN01–20LAA-1 (1949), 11; and AN01–20LAA-2, 94.
20. AN01–20LAA-2, 95.
21. AN01–20LAA-2, 95; and AN01–20LAA-1 (1949), 9.
22. AN01–20LAA-2, 95.

Chapter 8

1. *Lycoming O-290-7 Engine Specification No. 2028-A* (Williamsport, PA: Lycoming Division—The Aviation Corporation, August 9, 1945).
2. Wayne Dalrymple, *Organizational Requirements and Design Criteria Model XL-15* (Boeing Wichita Report No. WD-12110 Rev3, 14 March 1947), 46; *Airplane, Observation (Field Artillery) Construction Specification for Basic Flight Test Model XL-15, Boeing-Wichita Model m451, Contract W33–038 AC-15054, Airplane Serial No. 46-520* (Boeing Wichita Report No. WD-12102A, 29 August 1949), 26–27; *Handbook Erection and Maintenance Instructions USAF Model YL-15 Aircraft.* AN 01–20LAA-2 (10 January 1949), 1, 3, 13–14, 61–62, 65–72,94; and *Handbook Flight Operating Instructions USAF Model YL-15 Aircraft* (AN 01–20LAA-1, 23 March 1949), 1.
3. WD-12102A, 26; and AN01–20LAA-2, 3, 72.
4. *Model 451-2 Preliminary Performance, Stability and Control Characteristics* (Boeing-Wichita Report No. WD-12056, 6 March 1946), 6, 39–40; and AN01–20LAA-2, 73.
5. E. O. Weining, *Development of the Model L-15 Liaison Airplane and Design Development of Liaison Aircraft* (Boeing-Wichita Report No. WD-13024, 20 June 1950), 5; WD-12102A, 12,34; and AGF Air Training School Staff, "A.G.F. Light Aviation." *The Field Artillery Journal*, Vol. 36, No. 10, Oct 46, 585.
6. *Pilot's Handbook Flight Operating Instructions for XL-15 Airplane Serial Number 46-521* (Boeing Wichita Document WD-12174), 13.
7. Aircraft Project and Historical File 1942–1951 YL-15-47-423, RG 342, NARA II.

Chapter 9

1. E. O. Weining, *Development of the Model L-15 Liaison Airplane and Design Development of Liaison Aircraft* (Boeing-Wichita Report No. WD-13024, 20 June 1950), 15.
2. *Airplane, Observation (Field Artillery) Construction Specification for Basic Flight Test Model XL-15, Boeing-Wichita Model m451, Contract W33–038 AC-15054, Airplane Serial No. 46-520* (Boeing Wichita Report No. WD-12102A, 29 August 1949), 23–24; Wayne Dalrymple, *Organizational Requirements and Design Criteria Model XL-15* (Boeing Wichita Report No. WD-12110 Rev3, 14 March 1947), 38–39; *Handbook Erection and Maintenance Instructions USAF Model YL-15 Aircraft* (AN 01–20LAA-2, 10 January 1949), 47–61; and *Handbook Flight Operating Instructions USAF Model YL-15 Aircraft* (AN 01–20LAA-1, 23 March 1949), 48.
3. WD-12110, 40.
4. WD-12110, 39–40; and E. O. Weining, *Model XL-15 Float Installation Hydrodynamic and Structural Calculations* (Boeing-Wichita Report No. WD-12261, 31 March 1948), 4, 41, 47.
5. WD-12261, 4.
6. WD-12261, 41.
7. "WWII Brodie landing system for Army L-4," https://groups.google.com/forum/#!topic/rec.aviation.military/NeYTied2rZw (accessed July 6, 2017).

Chapter 10

1. *Handbook Erection and Maintenance Instructions USAF Model YL-15 Aircraft* (AN 01–20LAA-2,10 January 1949), 73–74.
2. AN01–20LAA-2, 71–72, 75; and *Handbook Flight Operating Instructions USAF Model YL-15 Aircraft* (AN 01–20LAA-1, 23 March 1949), 1.
3. AN01–20LAA-2, 74; and AN01–20LAA-1 (1949), 7.
4. AN01–20LAA-2, 92; and AN01–20LAA-1 (1949), 14.
5. AN01–20LAA-2, 73–74; and AN01–20LAA-1 (1949), 7.
6. AN01–20LAA-1 (1949), 7, 13.
7. AN01–20LAA-2, 29–30.
8. AN01–20LAA-2, 34.
9. AN01–20LAA-1 (1949), 2.
10. AN01–20LAA-2, 38; and AN01–20LAA-1 (1949), 2.
11. AN01–20LAA-2, 30.
12. AN01–20LAA-2, 94; and AN01–20LAA-1 (1949), 11.
13. AN01–20LAA-2, 33.
14. AN01–20LAA-2, 32.
15. AN01–20LAA-2, 36.
16. AN01–20LAA-2, 36.
17. AN01–20LAA-2, 40.
18. AN01–20LAA-1 (1949), 3–4.
19. AN01–20LAA-2, 40.
20. AN01–20LAA-2, 40; and AN01–20LAA-1 (1949), 4.
21. AN01–20LAA-2, 87, 118.
22. AN01–20LAA-2, 87.
23. AN01–20LAA-2, 88.
24. AN01–20LAA-2, 90–91; and AN01–20LAA-1 (1949), 27–28.
25. AN01–20LAA-1 (1949), 91,
26. AN01–20LAA-1 (1949), 91,
27. AN01–20LAA-1 (1949), 91; and *Parts*

Catalog USAF Model YL-15 Aircraft (AN 01–20LAA-4, 12 January 1949), 57.
28. AN01–20LAA-2, 91.
29. AN01–20LAA-2, 91.
30. AN01–20LAA-2, 91; and AN01–20LAA-1 (1 (1949), 28.
31. AN01–20LAA-2, 73; and AN01–20LAA-1 (1949), 8.
32. AN01–20LAA-2, 92.
33. AN01–20LAA-2, 92.
34. AN01–20LAA-2, 92; and AN01–20LAA-1 (1949), 28.
35. AN01–20LAA-2, 92.
36. AN01–20LAA-2, 92.
37. AN01–20LAA-2, 92; and AN 01–20LAA-4, 58.
38. AN01–20LAA-2, 76.
39. AN01–20LAA-2, 76, 84–85; and AN01–20LAA-1 (1949), 8–9.
40. AN01–20LAA-2, 80.
41. AN01–20LAA-2, 84–85.
42. AN01–20LAA-2, 81.
43. AN01–20LAA-2, 85.
44. AN01–20LAA-2, 16, 85.
45. AN01–20LAA-2, 85.
46. AN01–20LAA-2, 85, 112; and AN 01–20LAA-4, 59–60.
47. AN01–20LAA-2, 82.
48. AN01–20LAA-2, 84.
49. AN01–20LAA-2, 84.
50. AN01–20LAA-2, 82–83.
51. AN01–20LAA-2, 83.
52. AN01–20LAA-2, 82.
53. E.O. Weining, *Development of the Model L-15 Liaison Airplane and Design Development of Liaison Aircraft* (Boeing-Wichita Report No. WD-13024, 20 June 1950), 14.
54. AN01–20LAA-2, 85–86.
55. AN01–20LAA-2, 85–86.
56. AN01–20LAA-2, 86.
57. AN01–20LAA-2, 86.
58. AN01–20LAA-2, 86.
59. AN01–20LAA-2, 86.
60. AN01–20LAA-2, 86.
61. *Pilot's Handbook Flight Operating Instructions for XL-15 Airplane Serial Number 46–521* (Boeing Wichita Document WD-12174), 35; and WD-13024, 5; *Airplane, Observation (Field Artillery) Construction Specification for Basic Flight Test Model XL-15, Boeing-Wichita Model m451, Contract W33–038 AC-15054, Airplane Serial No. 46–520* (Boeing Wichita Report No. WD-12102A, 29 August 1949), 31–32.
62. AN01–20LAA-2, 94–95.
63. AN01–20LAA-2, 95.
64. AN01–20LAA-2, 46–47, 95; and AN01–20LAA-1 (1949), 24–25.

Chapter 11

1. Wayne Dalrymple, *Organizational Requirements and Design Criteria Model XL-15* (Boeing Wichita Report No. WD-12110 Rev3, 14 March 1947), 28
2. *Airplane, Observation (Field Artillery) Construction Specification for Basic Flight Test Model XL-15, Boeing-Wichita Model m451, Contract W33–038 AC-15054, Airplane Serial No. 46–520* (Boeing Wichita Report No. WD-12102A, 29 August 1949), 13–16.
3. Onya A. Kelley, *Model XL-15 Actual Weight and Balance* (Boeing-Wichita Report No. WD-10321, March 5, 1948), 5.
4. *Model 451-2 Preliminary Performance, Stability and Control Characteristics* (Boeing-Wichita Report No. WD-12056, 6 March 1946), 7, 18–20
5. James M. Wickham, *YL-15 Standard Characteristics* (Boeing Wichita Report No. WD-12366, 15 July 1949), 5.
6. WD-10321, 84.
7. WD-10321, 41, 68, 83.
8. E.O. Weining, *Liaison Airplane Weight and Performance Study* (Boeing-Wichita Report No. WD-12396, 22 December 1949), 7–33; and Jan Roskam, *Airplane Design—Part V: Component Weight Estimation*. Ottawa, Kansas: Roskam Aviation and Engineering Corp., 1989, 128–129.

Chapter 12

1. *Handbook Erection and Maintenance Instructions USAF Model YL-15 Aircraft*. AN 01–20LAA-2 (10 January 1949), 8–9.
2. AN01–20LAA-2, 9.
3. AN01–20LAA-2, 7–8.
4. AN01–20LAA-2, 8.
5. *Boeing Scout* (Boeing Wichita Document WD-12276).
6. AN01–20LAA-2, 9; and *Handbook Flight Operating Instructions USAF Model YL-15 Aircraft* (AN 01–20LAA-1, 23 March 1949), 4.
7. AN01–20LAA-2, 9.
8. AN01–20LAA-2, 10.
9. AN01–20LAA-2, 8–9.
10. *Pilot's Handbook Flight Operating Instructions for XL-15 Airplane Serial Number 46–521* (Boeing-Wichita Document WD-12174), 24.
11. AN01–20LAA-2, 8–9; *Parts Catalog USAF Model YL-15 Aircraft*. (AN 01–20LAA-4, 12 January 1949), 86; and WD-12174, 24.
12. WD-12174, 24.
13. WD-12174, 24.
14. AN01–20LAA-2, 8.
15. AN01–20LAA-2, 11.
16. AN01–20LAA-2, 11–13; and WD-12276.
17. AN01–20LAA-2, 11–13.
18. AN01–20LAA-2, 13.
19. AN01–20LAA-2, 13.
20. WD-12276; and E. O. Weining, *Airplane Characteristics Summary Model YL-15* (Boeing Wichita Report No. WD-12280, 27 May 1948), 4.
21. AN01–20LAA-2, 43–46, 50–56.

Chapter 13

1. E. H. Rowley, *Time Before Space: An Airman's Odyssey—from Biplanes to Rockets* (Manhattan, KS: Sunflower University Press, 1994), 223, 225, 227; and William D. Thompson, *A Test Pilot's Life* (Bend, OR: Maverick Publications, 1997), 20, 23.
2. D. C. Heimburger, *Pilot Comments and Flight Performance Data on the YL-15 Airplane.* (Boeing-Wichita Report No. WD-12827, 10 Jan 1949), 4.
3. WD-12827, 5.
4. WD-12827, 5–6.
5. WD-12827, 6.
6. WD-12827, 6–7.
7. WD-12827, 8–9.
8. WD-12827, 11–12, 20.
9. WD-12827, 9, 14–17, 25–26.
10. WD-12827, 13, 23.

Chapter 14

1. Steven G. Steffens, *Memorandum Report on Accelerated Service Test of YL-15 Aircraft.* MCRFT-2213 (Wright Field, Dayton, OH: Accelerated Service Test Branch, Air Materiel Command, 12 April 1949) 1.
2. MCRFT-2213, 3.
3. MCRFT-2213, 3.
4. MCRFT-2213, 4.
5. MCRFT-2213, 6.
6. MCRFT-2213, 6.
7. MCRFT-2213, 6.
8. MCRFT-2213, 7.
9. MCRFT-2213, 7–8.
10. MCRFT-2213, 8.
11. MCRFT-2213, 8.
12. MCRFT-2213, 8.
13. MCRFT-2213, 9.
14. MCRFT-2213, 9.
15. MCRFT-2213, 9.
16. MCRFT-2213, 9–11.

Chapter 15

1. Edgar F. Raines, Jr., *Eyes of Artillery—The Orgins of Modern U.S. Army Aviation in World War II.* Honolulu, HI: University Press of the Pacific, 2002; Robert S. Brown, Maj. USA, "The Development of Organic Light Aviation in the Army Ground Forces in World War II" (master's thesis, U.S. Army Command and General staff College, Fort Leavenworth, KS, 2000); Irving B. Holley, Jr, *Evolution of the Liaison-Type Airplane, 1917–1944,* Army Air Forces Historical Studies: No. 44, AAF Historical Office, Hq., AAF, April 1946; Kent Roberts Greenfield, Col, Inf. Res, *Army Ground Forces and the Air-Ground Battle Team Including Organic Light Aviation.* Army Ground Forces Historical Study No. 35, Historical Section AGF, 1948; Robert F. Futrell, *Command of Observation Aviation: A study in Control of Tactical Airpower.* USAF Historical Studies No. 24, USAF Historical Division, Air University, Maxwell AFB, AL, September 1956, 1–21; and Mal Holcomb, "Send Grasshopper." *Wings*, Vol. 13. No. 1, Feb 1983, 40–49.
2. AGF Hist. Study No. 35, 62–65,104–108; and Raines, 119–122, 286, 299–300.
3. Raines, *Eyes of Artillery,* 300, 331.
4. A.R. Crawford, Brig. Gen., USA, Chief, Research & Engineering Division, Hq, AAF, Washington D.C., "Mock-Up Inspection of XL15 Airplane," letter to Commanding General, Air Materiel Command, Wright Field, Dayton, OH, 21 August 1946.
5. AGF Air Training School Staff, "A.G.F. Light Aviation." *The Field Artillery Journal,* Vol. 36, No. 10, Oct 46, 582–585.
6. Herbert L. Earnest, Maj. Gen, Asst Chief of Staff G-3, Army Ground Forces, "Army Ground Forces Aircraft," Summary Sheet for the Chief of Staff, Army Ground Forces, 20 Dec 1946; and Charles L. Bolte, Maj. Gen, Chief of Staff, Army Ground Forces, "Army Ground Forces Aircraft," letter to Commanding General, Army Air Forces, Washington, D.C., 28 Dec 1946.
7. Ira G. Eaker, Lt Gen, USA, Deputy Commander, Army Air Forces, "Army Ground Forces Aircraft," letter to Commanding General, Army Ground Forces, Fort Monroe, VA, 14 Feb 1947; Herbert L. Earnest, Maj. Gen, Asst Chief of Staff G-3, Army Ground Forces, "Army Ground Forces Aircraft," Summary Sheet for the Chief of Staff, Army Ground Forces, 11 April 1947; Robert M. Bathurst, Brig. Gen., Deputy Chief of Staff, Army Ground Forces, "Army Ground Forces Aircraft," "Army Ground Forces Aircraft," letter to Commanding General, Army Air Forces, Washington, D.C., 17 April 1947; and Director of Statistical Services. *United States Air Force Statistical Digest 1947.* Comptroller, HQ, USAF, Washington, D.C., August 1948, 114.
8. George F. Smith, Col, Air Corps, Chief Aircraft Project Section, Engineering Division, Air Materiel Command, Wright Field, Dayton, OH, "Acceptance Inspection of the XL-15 Airplane," routing and record sheet to TSENG and TSEOA-5, 16 Sept 1947.
9. Richard L. Long, Major, USAF, AGF Liaison Officer for Light Aviation, Air Materiel Command, Wright Field, Dayton, OH, "Authority to Fly the XL-15 Airplane," routing and record sheet to MCREOA—Mr. L.V. Cook, 13 Jan 1948; George F. Smith, Col, USAF, Chief Aircraft Project Section, teletype message to USAF Officer in Charge, Boeing Airplane Company, Wichita Division, Wichita, Kansas, 15 Jan 1948 :and George F. Smith, Col, Air Corps, Chief Aircraft Project Section, teletype message to USAF Officer in Charge, Boeing Airplane Company, Wichita Division, Wichita, Kansas, 23 Jan 1948.
10. George F. Smith, Col, USAF, Chief Aircraft Projects Section, Engineering Division," Contract W33–038 ac-15054, Model XL-15 Airplane, Performance Guarantees" letter to AFMRD, Deputy

Chief of Staff, Material, USAF, Washington, D.C., 4 February 1948.

11. George F. Smith, Col, USAF, Chief Aircraft Projects Section, Engineering Division," Contract W33–038 ac-15054, Model XL-15 Airplane," letter to Mr. Suchy MCPPXA-5, 24 Mar 1948.

12. George F. Smith, Col, USAF, Chief, Aircraft Project Section, teletype message to USAF Officer in Charge, Boeing Airplane Company, Wichita Division, Wichita, Kansas, 26 Mar 1948; and L. V. Cook, *XL-15 Liaison Airplane, "Scout": Final Report of Development, Procurement, Inspection, Testing, and Acceptance,* AF Technical Report No. 6046 (Wright-Patterson Air Force Base, Dayton, OH: USAF Air Materiel Command, July 1949), 5

13. H. A. Craig, Lt. Gen USAF, Deputy Chief of Staff, Material, HQ USAF, Washington, D.C., XL-15 and YL-15," letter to Commanding General, Air Materiel Command, Wright-Patterson AFB, Dayton, OH. Attn: MCREOA, 28 April 1948; and *USAF Statistical Digest 1947,* 114.

14. George E. Schaetzel, Col, USAF, Chief, Aircraft & Missiles Section, Engineering Division, "Installation of 140 H.P. Engine in XL-15 Airplane" conference minutes 5 Nov 1948.

15. Aircraft Project and Historical File 1942–1951 YL-15-47-423, RG 342, NARA II; Boeing Press Release W-414A.

16. Boeing Press Release W-414A.

17. William D. Thompson, *Cessna—Wings for the World: The Single-Engine Development Story* (Bend, OR: Maverick Publications, 1991), 174–176; and Jeffrey L Rodengen, *The Legend of Cessna* (Fort Lauderdale, FL: Write Stuff Enterprises, Inc., 1997), 108–111.

Chapter 16

1. See the section "Looks at More Power" in Chapter 6, "Performance."

2. E. O. Weining, Project Engineer, "Model L-15—Development of," Inter Office Memo to N.D. Showalter, Boeing-Wichita, 25 July 1950.

3. E. O. Weining, Project Engineer, "Model XL-15—Design Development of," Inter Office Memo to Harold W. Zipp, Boeing-Wichita, 13 March 1950; E. O. Weining, Project Engineer, "Model L-15 Design Changes," Inter Office Memo to Harold W. Zipp, Boeing-Wichita, 19 June 1950; and E. O. Weining, Project Engineer, "Model L-15—Development of," Inter Office Memo to N.D. Showalter, Boeing-Wichita, 25 July 1950.

4. "Utility Airplane Prospectus Summary," undated paper believed to have been presented to the Flying Farmers.

Bibliography

Books

Bowers, Peter M. *Unconventional Aircraft*. Blue Ridge Summit, PA: Tab Books Inc., 1984.

Civil Aviation and the National Economy. Washington, D.C.: United States Dept. of Commerce, Civil Aeronautics Administration, Sept 1945.

Francillon, Rene J. *Lockheed Aircraft Since 1913*. Annapolis, MD: Naval Institute Press, 1988.

Francillon, Rene J. *McDonnell Douglas Aircraft Since 1920: Volume I*. Annapolis, MD: Naval Institute Press, 1988.

Juptner, Joseph P. *U.S. Civil Aircraft, Volume 8*. Fallbrook, CA: Aero Publishers Inc., 1980

Mertens, Randy. *Closet Cases*. Kansas City, MO: Pilot News Press, 1980.

Phillips, Edward H. *Stearman Aircraft: A Detailed History*. North Branch, MN: Specialty Press, 2006.

Raines, Edgar F., Jr. *Eyes of Artillery—The Origins of Modern U.S. Army Aviation in World War II*. Honolulu, HI: University Press of the Pacific, 2002.

Rodengen, Jeffrey L. *The Legend of Cessna*. Fort Lauderdale, FL: Write Stuff Enterprises, Inc., 1997.

Roskam, Jan. *Airplane Design—Part V: Component Weight Estimation*. Ottawa, KS: Roskam Aviation and Engineering Corp., 1989.

Rowley, E. H. *Time Before Space: An Airman's Odyssey—from Biplanes to Rockets*. Manhattan, KS: Sunflower University Press, 1994.

Simpson, Rod. *The General Aviation Handbook*. Hinckley, England: Midland Publishing, 2005.

Thompson, William D. *Cessna—Wings for the World: The Single-Engine Development Story*. Bend, OR: Maverick Publications, 1991.

Thompson, William D. *A Test Pilot's Life*. Bend, OR: Maverick Publications, 1997.

Wagner, Ray. *Mustang Designer*. New York: Orion Books, 1990.

Correspondence

Bathurst, Robert M., Brig. Gen., Deputy Chief of Staff, Army Ground Forces, "Army Ground Forces Aircraft," "Army Ground Forces Aircraft," letter to Commanding General, Army Air Forces, Washington, D.C., 17 April 1947.

Bolte, Charles L., Maj. Gen, Chief of Staff, Army Ground Forces, "Army Ground Forces Aircraft," letter to Commanding General, Army Air Forces, Washington, D.C., 28 Dec 1946.

Commanding General, Air Materiel Command, "Informal Design Competition—Field Artillery Observation Aircraft (Liaison Type)," letter to Commanding General. Army Air Forces, Washington, D.C., 18 April 1946.

Craig, H. A., Lt. Gen USAF, Deputy Chief of Staff, Material, HQ USAF, Washington, D.C., XL-15 and YL-15," letter to Commanding General, Air Materiel Command, Wright-Patterson AFB, Dayton, OH, Attn: MCREOA, 28 April 1948.

Crawford, A. R., Brig. Gen., USA, Chief, Research & Engineering Division, Hq, AAF, Washington, D.C., "Mock-Up Inspection of XL15 Airplane" letter to Commanding General, Air Materiel Command, Wright Field, Dayton, OH, 21 August 1946.

Davis, R. HB., Major, Air Corps, Acting Chief of Administration, Air Technical Service Command, Wright Field, Dayton, OH, "Field Artillery Observation Aircraft," Technical Instructions TI -2133, Addendum No. 3 to Engineering Division, 27 September 1945.

Eaker, Ira G., Lt Gen, USA, Deputy Commander, Army Air Forces, "Army Ground Forces Aircraft," letter to Commanding General, Army Ground Forces, Fort Monroe, VA, 14 Feb 1947.

Earnest, Herbert L., Maj. Gen, Asst Chief of Staff G-3, Army Ground Forces, "Army Ground Forces Aircraft," Summary Sheet for the Chief of Staff, Army Ground Forces, 20 Dec 1946.

Earnest, Herbert L., Maj. Gen, Asst Chief of Staff G-3, Army Ground Forces, "Army Ground Forces Aircraft," Summary Sheet for the Chief of Staff, Army Ground Forces, 11 April 1947.

Long, Richard L., Major, USAF, AGF Liaison Officer for Light Aviation, Air Materiel Command, Wright Field, Dayton, OH, "Authority to Fly the XL-15 Airplane," routing and record sheet to MCREOA—Mr. L. V. Cook, 13 Jan 1948.

Price, George E., Col, Air Corps, Chief Aircraft Projects Section, Service Engineering Subdivision, Engineering Division, Air Technical Service Command, Wright Field, Dayton, OH, "Field Artillery Observation Aircraft—Request for Bids," letter to 26 aircraft manufacturers, 4 February 1946.

Schaefer, J. E., Vice President, Wichita Division, Boeing Airplane Company, "Field Artillery Observation Aircraft—Request for Bids," letter to Commanding General, Air Technical Service Command, Wright Field, Dayton, OH, 4 March 1946.

Schaefer, J. E., Vice President, Wichita Division, Boeing Airplane Company, "Contract W33-038 ac-15054, Model XL-15 Airplane—Alternate Engine Installation," letter to Commanding General, Air Materiel Command, Wright Field, Dayton, OH, 7 July 1947.

Schaefer, J. E., Vice President, Wichita Division, Boeing Airplane Company, "Contract W33-038 ac-16945, Model YL-15 Airplanes—Proposal for Engine Change," letter to Commanding General, Air Materiel Command, Wright Field, Dayton, OH, 15 October 1947.

Schaetzel, George E., Col, Air Corps, Chief, Aircraft Section, Procurement Division, Air Materiel Command, "W33-038 ac-16945, Model YL-15 Aircraft Revision of Planning" letter to Chief of Staff US Air Force, Washington, D.C., 24 October 1947.

Schaetzel, George E., Col, USAF, Chief, Aircraft Section, Procurement Division, Air Materiel Command, "W33-038 ac-16945, Model YL-15 Aircraft Revision of Planning" letter to USAF Inspector in Charge, Boeing Airplane Company, Wichita Division, Wichita, KS, 5 December 1947.

Smith, George F., Col, Air Corps, Chief Aircraft Project Section, Engineering Division, Air Materiel Command, Wright Field, Dayton, OH, "Acceptance Inspection of the XL-15 Airplane," routing and record sheet to TSENG and TSEOA-5, 16 Sept 1947.

Smith, George F., Col, USAF, Chief Aircraft Project Section, teletype message to USAF Officer in Charge, Boeing Airplane Company, Wichita Division, Wichita, KS, 15 Jan 1948.

Smith, George F., Col, Air Corps, Chief Aircraft Project Section, teletype message to USAF Officer in Charge, Boeing Airplane Company, Wichita Division, Wichita, KS, 23 Jan 1948.

Smith, George F., Col, USAF, Chief Aircraft Projects Section, Engineering Division," Contract W33-038 ac-15054, Model XL-15 Airplane, Performance Guarantees," letter to AFMRD, Deputy Chief of Staff, Material, USAF, Washington, D.C., 4 February 1948.

Smith, George F., Col, USAF, Chief, Aircraft Projects Section, Engineering Division," Contract W33-038 ac-15054, Model XL-15 Airplane" letter to Mr. Suchy MCPPXA-5, 24 Mar 1948.

Smith, George F., Col, USAF, Chief, Aircraft Project Section, teletype message to USAF Officer in Charge, Boeing Airplane Company, Wichita Division, Wichita, KS, 26 Mar 1948.

Taylor, R. K., Brig. Gen., USA, Air Materiel Command, "Procurement of Observation Aircraft (Liaison Type)," Technical Instructions to Engineering Division, 6 May 1946.

Vandegrift, R. T., Service Manager, Lycoming Division, Avco Manufacturing Corp, Williamsport, PA, "Loan of O-290-A2 Engine for XL-15 Airplane" letter to Commanding General, Air Materiel Command, Wright Patterson AFB, Dayton, OH, Attn: MCREOA, December 14, 1948

Weining, E. O., Project Engineer. "Model XL-15—Design Development of," Inter Office Memo to Harold W. Zipp, Boeing-Wichita, 13 March 1950.

Weining, E. O., Project Engineer. "Model L-15 Design Changes," Inter Office Memo to Harold W. Zipp, Boeing-Wichita, 19 June 1950.

Weining, E. O., Project Engineer. "Model L-15 - Development of," Inter Office Memo to N. D. Showalter, Boeing-Wichita, 25 July 1950.

Zipp, Harold W., Chief Engineer, Wichita Division, Boeing Airplane Company, "AAF Liaison Airplane—Proposed Design For," letter to Commanding General, Air Technical Service Command, Wright Field, Dayton, OH, Attn: Engineering Division—Brig. Gen L. C. Craigie, 5 December 1945.

Zipp, Harold W., Chief Engineer, Wichita Division, Boeing Airplane Company, "Contract W33-038 ac-15054—Model XL-15 Airplane Fuel Requirements, Possible Engine Development Resulting from Change in," letter to Mr. J. B. Risbel, Chief Aircraft Project Section, Service Engineering Subdivision, Engineering Division, Air Materiel Command, Wright Field, Dayton, OH, 15 August 1946.

Zipp, Harold W., Chief Engineer, Wichita Division, Boeing Airplane Company, "Contract W33-038 ac-15054—Model XL-15 Airplane—Propeller for," letter to Commanding General, Air Materiel Command, Wright Field, Dayton OH, Attn: TSEOA-5, 18 March 1947.

Zipp, Harold W., Chief Engineer, Wichita Division, Boeing Airplane Company, "Contract W33-038 ac-15054—Model XL-15 Airplane—Extension of Contract Delivery Date (Change No. 19) and Request for Additional Flight Test Time (Change No. 20)," letter to Commanding General, Air Materiel Command, Wright Field, Dayton, OH, Attn: TSEOA-5, 20 July 1947.

Zipp, Harold W., Chief Engineer, Wichita Division, Boeing Airplane Company, "Contract W33-038 ac-15054—Model XL-15 Airplane—Accident Report on," letter to Commanding General, Air Materiel Command, Wright Field, Dayton, OH, Attn: TSEOA-5, 13 August 1947.

Manuals

Handbook Erection and Maintenance Instructions USAF Model YL-15 Aircraft. AN 01–20LAA-2, 10 January 1949.

Handbook Flight Operating Instructions USAF Model YL-15 Aircraft. AN 01–20LAA-1, 23 March 1949.

Parts Catalog USAF Model YL-15 Aircraft. AN 01-20LAA-4, 12 January 1949.

Pilot's Handbook Flight Operating Instructions for XL-15 Airplane Serial Number 46-521. Boeing-Wichita Document WD-12174.

Pilot's Handbook Flight Operating Instructions for YL-15 Airplane Serial Numbers 47-423 Through 47-432. AN 01-20LAA-1, Nov 1, 1948.

Papers, Etc.

Brown, Robert S., Maj. USA. "The Development of Organic Light Aviation in the Army Ground Forces in World War II." Master's thesis, U.S. Army Command and General staff College, Fort Leavenworth, KS, 2000.

Discussion of Engineering and Suitability Characteristics—Manufacturer's Ratings

Method of Evaluation for Aircraft, Observation (Field Artillery) Inclosure No. 3 of RFB

Proposal for Army Air Forces Observation Airplane. Consolidated Vultee Aircraft Corp., San Diego, CA, March 6, 1946.

Weining, E. O. "External Airfoil Flaps as a High-Lift Device," 29 Nov 1948.

Wickham, James M. "Research in Simplified Control," read by D. C. Heinburger to the Wichita Chapter Society of Automotive Engineers, 20 January 1949.

Periodicals

AGF Air Training School Staff. "A.G.F. Light Aviation." *The Field Artillery Journal*, Vol. 36, No. 10, Oct 46, 582-585.

Holcomb, Mal. "Send Grasshopper." *Wings*, Vol. 13. No. 1, Feb 1983, 40-49.

Spatt, George. "Wing on a Pivot." *Flying*, Vol. 38, No. 5, May 1946, 42-44, 112.

Reports

Airplane, Observation (Field Artillery) Construction Specification for Basic Flight Test Model XL-15, Boeing-Wichita Model XL-15, Airplane Serial No. 46-520, Boeing Wichita Report No. WD-12102, 1 October 1946.

Airplane, Observation (Field Artillery) Construction Specification for Basic Flight Test Model XL-15, Boeing-Wichita Model m451, Contract W33-038 AC-15054, Airplane Serial No. 46-520, Boeing-Wichita Report No. WD-12102A, 29 August 1949.

Basow, H.M. & B.C. Hainline. *Development Tests of Beaded Wing Skin*. Boeing-Wichita Report No. WD-10541, August 27, 1946.

Blake, C., F. Leward, and K. Smith, *Aerodynamic Report for Army Air Forces Observation Airplane*. Consolidated Vultee Aircraft Corp Report ZA-152, San Diego, CA, March 1, 1946.

Boeing Scout, Boeing-Wichita Document WD-12276.

Borgstrom, Thomas J., and Robert L. Northup, 1st Lt, USAF. *Memorandum Report on Comparative Propeller, Engine Cooling and Longitudinal Stability and Control on the XL-15, USAF No. 46-520*. MCRFT-2152, Wright-Patterson Air Force Base, Dayton, OH: Air Materiel Command, 20 July 1948.

Brown, W.F. *Preliminary Model Specification for Army Air Forces Observation Airplane (Field Artillery) One Lycoming O-290-7 Engine*. Consolidated Vultee Aircraft Corp Report ZD-016, San Diego, CA, 25 February 1946.

Cook, L.V. *XL-15 Liaison Airplane, "Scout": Final Report of Development, Procurement, Inspection, Testing, and Acceptance*. AF Technical Report No. 6046, Wright-Patterson Air Force Base, Dayton, OH: USAF Air Materiel Command, July 1949.

Dalrymple, Wayne. *Organizational Requirements and Design Criteria Model XL-15*. Boeing-Wichita Report No. WD-12110 Rev3, 14 March 1947.

De Remer J.E. *Memorandum Report on Mock-Up Inspection of Xl-15 Airplane at Boeing Aircraft Company, Wichita, Kansas on 5 September 1946*. TSEPP-436-14, Wright Field, Dayton, OH: Power Plant Laboratory, Engineering Division, Air Materiel Command, Army Air Forces, 18 October 1946.

Director of Statistical Services. *United States Air Force Statistical Digest 1947*. Comptroller, HQ, USAF, Washington, D.C., August 1948.

Futrell, Robert F. *Command of Observation Aviation: A Study in Control of Tactical Airpower*. USAF Historical Studies No. 24, USAF Historical Division, Air University, Maxwell AFB, AL, September 1956.

Greenfield, Kent Roberts, Col. Inf. Res. *Army Ground Forces and the Air-Ground Battle Team Including Organic Light Aviation*. Army Ground Forces Historical Study No. 35, Historical Section AGF, 1948.

Heimburger, D.C. *Pilot Comments and Flight Performance Data on the YL-15 Airplane*. Boeing-Wichita Report No. WD-12827, 10 Jan 1949.

Holley, Irving B., Jr. *Evolution of the Liaison-Type Airplane, 1917-1944*. Army Air Forces Historical Studies: No. 44, AAF Historical Office, Hq., AAF, April 1946.

Kelley, Onya A. *Model XL-15 Actual Weight and Balance*. Boeing-Wichita Report No. WD-10321, March 5, 1948.

King, Jack H., and Thomas J. Cecil, Capt, USAF. *Memorandum Report on Performance Tests of the XL-15 Airplane, USAF No. 46-521 with a Lycoming O-290-A2 140 Horsepower Engine Installed*. MCRFT-2243, Wright-Patterson Air Force Base, Dayton, OH: Air Materiel Command, 1 September 1949.

King, Jack H., Thomas J. Cecil, Capt, USAF, and John F. Conn, Capt, USAF. *Memorandum Report on Longitudinal Stability and Control*

Tests on the XL-15 Airplane, USAF No. 46–521, and the YL-15 Airplane, USAF NO. 47–423. MCRFT-2246, Wright-Patterson Air Force Base, Dayton, OH: Air Materiel Command, 1 September 1949.

La Fountain, Charles H. *Memorandum Report on Flight Tests of the XL-15, USAF No. 46–520.* MCRFT-2120, Wright-Patterson Air Force Base, Dayton, OH: Air Materiel Command, 5 January 1948.

La Fountain, Charles H and Robert H. Shaefer, LTC, USAF. *Memorandum Report on Flight Tests of the XL-15, USAF No. 46–520.* MCRFT-2120 (Addendum 1), Wright-Patterson Air Force Base, Dayton, OH: Air Materiel Command, 6 February 1948.

Lycoming O-290-7 Engine Specification No. 2028-A. Williamsport, PA: Lycoming Division—The Aviation Corporation, August 9, 1945.

Miller, N. *Wing Stress Analysis—Volume II for Models XL-15 & YL-15.* Boeing-Wichita Report No. WD-10666 Rev2, December 15, 1948.

Miller, Manley, & Morley. *Control Surface Analysis for Model XL-15 & YL-15.* Boeing-Wichita Report No. WD-10669 Rev2, December 15, 1948.

Model 451-2 Preliminary Performance, Stability and Control Characteristics. Boeing-Wichita Report No. WD-12056, 6 March 1946.

Piper Model PA-9 Engineering Data. Piper Aircraft Corp. Report No. 504, Lock Haven, PA, 1 March 1946.

Proposal for the Development of a Field Artillery Observation Airplane., Ludington-Griswold, Inc., Saybrook, CT, 6 March 1946.

Rowley, E. H. *Initial Ground and Flight Tests on XL-15.*

Schaetzel George E., Col, USAF, Chief, Aircraft & Missiles Section, Engineering Division, "Installation of 140 H.P. Engine in XL-15 Airplane" conference minutes, 5 Nov 1948.

Seibel, C. M. *Model XL-15 Estimated Performance, Stability and Control Characteristics.* Boeing-Wichita Report No. WD-12130, 12 June 1947.

Steffens, Steven G. *Memorandum Report on Accelerated Service Test of YL-15 Aircraft.* MCRFT-2213, Wright Field, Dayton, OH: Accelerated Service Test Branch, Air Materiel Command, 12 April 1949.

Type Specification for Land Observation Airplane. Bellanca Aircraft Corporation Specification A-2, February 28, 1946.

Type Specification for Land Observation Airplane. Bellanca Aircraft Corporation Specification A-3, February 28, 1946.

Wardle, M. C. *Aerodynamics Report, Preliminary Design of Liaison Observation Airplane.* Bellanca Aircraft Corporation Report No. 469, March 4, 1946.

Weining, E. O. *Flight Test of Carburetor with Automatic Altitude Compensation Model Xl-15.* Boeing Wichita Report No. WD-12254, 24 February 1948.

Weining, E. O. *Model XL-15 Float Installation Hydrodynamic and Structural Calculations.* Boeing-Wichita Report No. WD-12261, 31 March 1948.

Weining, E. O. *Summary Analysis of Phase I Flight Tests Model XL-15 Airplane.* Boeing Wichita Report No. WD-12129, 8 April 1948.

Weining, E. O. *Airplane Characteristics Summary Model YL-15.* Boeing-Wichita Report No. WD-12280, 27 May 1948.

Weining, E. O. *Cruise Control Model YL-15.* Boeing-Wichita Report No. WD-12321 Rev1, March 10, 1949.

Weining, E. O. *Liaison Airplane Weight and Performance Study.* Boeing-Wichita Report No. WD-12396, 22 December 1949.

Weining, E. O. *Development of the Model L-15 Liaison Airplane and Design Development of Liaison Aircraft.* Boeing-Wichita Report No. WD-13024, 20 June 1950.

Wickham, James M. *YL-15 Standard Characteristics.* Boeing-Wichita Report No. WD-12366, 15 July 1949.

Wickham, Rowley, and E.O. Weining. *Summary of Phase III Flight Tests Model XL-15 Airplane.* Boeing Wichita Report No. WD-12303, 17 December 1948.

Websites

"Boeing Wichita History," https://wingsoverkansas.com/legacy/a375/ (accessed Mar 30, 2018).

"Ludington-Griswold," http://aerofiles.com/_lo.html (accessed Nov 13, 2018).

Wikipedia contributors, "Convair Model 118," *Wikipedia, The Free Encyclopedia,* https://en.wikipedia.org/wiki/Convair_Model_118 (accessed May 17, 2017).

"WWII Brodie landing system for Army L-4," https://groups.google.com/forum/#!topic/rec.aviation.military/NeYTied2rZw (accessed July 6, 2017).

Index

Numbers in **_bold italics_** indicate pages with illustrations

accelerated service test 209–216
accident 80
Aeronca 7
Aeronca L-16 Champion 1, 324
AGF flying the XL-15 225, 226
air tow 32, **_59_**, 136, 137
Aircraft Engine History Society 1
Army Air Forces liaison squadrons 218, 219, 221
Army Ground Forces get liaison aircraft 175, 217–221
Army National Guard 1, 224, 225
Aviation Week 1

Baker, Don 145
Beech Aircraft 1, 235
Bellanca 33–34, 38
Bellanca model 20-13 34, **_35_**, 38, 48, 49, 50
Bellanca model 24-19 34, **_36_**, 38, 48, 49, 50
Bellanca YO-50 34
Boeing B-29 Superfortress 3, **_178_**
Boeing B-47 Stratojet 27, 28, 194, 234
Boeing C-97 190, **_192_**, **_193_**
Boeing comments on flying the airplane 194–208
Boeing Model 451-2 37, 38–40, 47 **_48_**, **_49_**, 50, 52, 54–63, **_55_**, **_59_**, **_61_**
Boeing Model 451-4a 102
Bolte, Charles L. 222, 224
Borgstrom, Thomas J. 99
Bristol, D. L. 222
Brodie gear 30, 60, **_61_**, 89, 143, **_144_**, 145
Brown, Bill 1

Ceicl, Thomas J. 104
Cessna 7, **_8_**, 235
Cessna 170 230
Cessna 195 230
Cessna 305 *see* Cessna L-19
Cessna L-19 Bird Dog 1, 182
Cessna loans Ercoupe 24
Civil Air Patrol 1
Cogswell, David G. 225, 226
Collier's magazine 7
Colonial Skimmer 12
Consolidated Vultee (Convair)

proposal **_38_**, **_39_**, 40–42, 48, 49, 50
Continental 100 hp engine 10
Continental 250 hp engine 9
Continental C-145 102, 103, 104
Continental E-165 47
Continental O-290-11 230
control cable disconnect fittings **_155_** Convair 12–15
Convair L-13A 182
Convair Model 103 Skycar 13, **_14_**, 15, 20, **_21_**
Convair Model 116 ConvAirCar 13
Convair Model 118 ConvAirCar 13, 14, **_15_**
Crosby 26 hp engine 13
Curtiss O-1 175

data case 122, **_123_**
Douglas Model 1015 Cloudster II 9, **_10_**
Douglas O-46 175
Douglas XB-42 Mixmaster 9

Eaker, Ira C. 224
Earnest, J.G. 222
Eisenhower, Dwight D. 76–77, 92, 97, 220
electrical system 164–172, **_173_**
elevator control system 149, **_150_**, 151
elevator trim tab control system 151, 152
engine installation 126–135
engineering acceptance of first XL-15 225
equipment weights, optional 182
Ercoupe 1, 16, 20, 24, **_25_**
Evans, B. 222

Fairchild C-82 99, 190, **_192_**
Farrell, F.W. 225
The Field Artillery Journal 1, 222
fire extinguisher 123, **_124_**
first-aid kit 122
first flight 75, 76, 97
flap 18–25, 55, 66, 86, 87, 108, 109
flap control system 156, 157, **_158_**
Fleck, Wilbur E. 97

flight report and map case 119, 122
flight test instrumentation 82, 83, **_84_**, **_85_**, **_86_**
floats 30, 60, 141, **_142_**, 143, **_144_**
follow-on proposed liaison projects 231, 232, 233
Ford, W.W. 222
Fort Bragg, NC 209, 227, 228, 229
Fort Campbell, Kentucky 99
Fort Riley, Kansas 76
Franklin 4A4 engine 13
Franklin 50 hp engine 10
Franklin engine 8
Franklin O-335 102, 103
fuel system 146–148
fuselage 58, **_59_**, 111–118

General Skyfarer 16, **_17_**, 20
Griswold, Roger W., II **_41_**, 42, 44
Griswold's Integrated Wing **_41_**, 42, 44
group weight statement 178, 179
Grumman F4F Wildcat 12
Grumman G-63 Kitten 12
Grumman G-65 Tadpole 12, 14
Grumman G-72 Kitten II 12, **_13_**

hand towing 187
heating and ventilation 173, 174
Heimburger, Doug C. 101, 194, 197, 199, 204
Hogg, J. 221
hoisting the airplane 183, 184
hoisting the engine 184, **_185_**
horizontal tail 56, 57, 65, 110, 111

Institute of Aeronautical Sciences 101
instruments 159, 160, **_161_**, 162, 163, 164
Interstate L-6 22–28, 67–68, 182

jacking 183
Jackson, Paul E. 226
JATO 136, 147
jumped vertically 197

Index

Kelly, Onya A. 176
King, Jack H. 104
Kipfer, A.D. 176

L-19 fly-off competition 229, 230, 231
La Fountain, Charles H. 97, 98
landing gear 60, 64, 138–141
Leadville, Colorado 229
Lear 235
LeFever, C.W. 222, 225
leveling 183
Lindbergh, Charles 89, 92
Lockheed Model 33 Little Dipper 10, 11, 12
Lockheed Model 34 Big Dipper 10, 11, 12
Long, Richard L. 225, 226
Ludington, Charles Townsend 42
Ludington-Griswold, Inc. 42–44
Ludington-Griswold XLG-3 *41, 42*
Ludington-Griswold XLG-31 42, 43, 44, 48, 49, 50, 52
Lycoming GO-290 44
Lycoming GO-290-A 102
Lycoming O-290-7 40, 41, 47, 49, 59, 102, 126, 204, 205, 220; power curve 127
Lycoming O-290-A2 *103*, 104, 105, *231*
Lycoming O-435 102
Lycoming O-435-11 47, 49
Lycoming O-435C engine 13

maintenance and repair 190, 191. *193*
Marinelli, Jack 1, 222, 225, 226
McCutcheon, Kimble D. 2
message containers and bags 123
message pick-up equipment 122
mockup 68, *69, 70, 71, 72*, 221, 222

National Archives 1
North American Navion 9, *10*
North American O-47 175
North American P-51 Mustang 9, *10*
Northrup, Robert L. 99

off the shelf purchase 224, 225

parking 185
parking harness 118, *119*

performance guarantees, failure to meet 226
Phase I flight tests 73, 75–82
Phase II flight tests 97
Phase III flight tests 84–89
Phase IV flight tests 99
Pierce, Wilfred (Bill) A. 1, 106
Piper 7
Piper J-3 Cub 1, 3, *217*
Piper L-4 1, *45*, 143, 145, 182, *218*
Piper L-4X 220
Piper PA-9 3, 44, *46* 47, 48, 49, 50, *51, 52*
Piper PA-28 Cherokee 11
Piper YL-14 44, *45, 220*
pitot-static system 158, 159
Pratt-Read PR-G1 (LNE-1, TG-32) 42
preparation for transport 187, 188, 189, *190*
propeller 40, 59, 60, 68, 79, 80, 97, 98, 99, 102, 104, 130, 135, 136, 186, 203
proposed utility airplane 234
pyrotechnic projector 124, *125*

rear vision mirror 123
relief tubes 119, *120, 121*
Republic F-84 Thunderjet 9
Republic Seabee 9
Republic Thunderbolt Amphibian 8
Request for Bids 3, 29–33
Riley, D.E. 222
Rowley, Elton H. 26, 64, 75–77, 92, 97, 101
Ruckman, T.M. 222
rudder control system 153, 154, *156*
Ryan Aeronautical Navion 9

seats 119, *120*
Seibel, Charles M 96
service test 104, 209, 224, 227–229
Shaefer, Robert R. 98
Sheppard, C.L. 222, 225
signal flare container 125
simplified control 16, 18, 20, 27
skis 30, 60, *61*, 141, *142*
Smith, L.B. 225
Spaatz, Carl 52
spoiler ailerons 55, 56, 66, 109, 110; control system 155, *157*

Spratt, George 13
Spratt controllable wing concept 13, *14, 21*
stall/spin tests 77–79, 86, 88, 89
station lines 181
Stearman-Hammond 16, *18*, 20
Stearman trainers 3
Stearman XOSS-1 19
Stinson Flying Station Wagon 13
Stinson L-1 *see* Stinson O-49
Stinson L-5 Sentinel 1, 143, 218, 219
Stinson O-49 34, 175
Stinson Voyager 13
Stinson XL-13 42
structural integrity flight tests 82, 88, 205
structural load tests 92, 226
structural tests of wing panels 64, *65*

tail wheel control system 154, 155, *156*
Taylorcraft 7
Thorp, John 11
Thurston, David 12
tie-down 186, *187*
towing by jeep or truck 189, *191*
transporting by cargo airplane 99, 189, 190
transporting on two & one-half ton truck 31, 189. *191*
Truman, Harry S. 229

U.S. Fish & Wildlife Service 230, 235
University of Wichita wind tunnel 21, *23*

vertical tail 58, 65, 111

weather protection covers 186, *188*
weight effect on performance 176
Weining, Earl O. 1, 28, 54, 100, 231, 233, 234
Wickham, James M. 17, 28, 101
wind tunnel testing 20–23, 64–67, 233, 234
wing 55, 56, 106–108

zero-thrust indicator 83, *85*
Zipp, Harold 77

www.ingramcontent.com/pod-product-compliance
Lightning Source LLC
Chambersburg PA
CBHW060339010526
44117CB00017B/2887